The Metaphysics of Time

New Dialogues in Philosophy

A series in dialogue form, explicating foundational problems in the philosophy of existence, knowledge, and value

Series Editor
Professor Dale Jacquette, Senior Professorial Chair in Theoretical Philosophy, University of Bern, Switzerland

In the tradition of Plato, Berkeley, Hume, and other great philosophical dramatists, Rowman & Littlefield presents an exciting new series of philosophical dialogues. This innovative series has been conceived to encourage a deeper understanding of philosophy through the literary device of lively argument in scripted dialogues, a pedagogic method that is proven effective in helping students to understand challenging concepts while demonstrating the merits and shortcomings of philosophical positions displaying a wide variety of structure and content. Each volume is compact and affordable, written by a respected scholar whose expertise informs each dialogue, and presents a range of positions through its characters' voices that will resonate with students' interests while encouraging them to engage in philosophical dialogue themselves.

Titles

J. Kellenberger, *Moral Relativism: A Dialogue* (2008)
Michael Ruse, *Evolution and Religion: A Dialogue* (2008)
Charles Taliaferro, *Dialogues about God* (2008)
Brian Orend, *On War: A Dialogue* (2008)
Dale Jacquette, *Dialogues on the Ethics of Capital Punishment* (2009)
Bradley Dowden, *The Metaphysics of Time: A Dialogue* (2009)

Forthcoming titles

Michael Krausz, *Relativism: A Dialogue*
Dan Lloyd, *Ghosts in the Machine: A Dialogue*

The Metaphysics of Time

A Dialogue

Bradley Dowden

ROWMAN & LITTLEFIELD PUBLISHERS, INC.
Lanham • Boulder • New York • Toronto • Plymouth, UK

ROWMAN & LITTLEFIELD PUBLISHERS, INC.

Published in the United States of America
by Rowman & Littlefield Publishers, Inc.
A wholly owned subsidiary of The Rowman & Littlefield Publishing Group, Inc.
4501 Forbes Boulevard, Suite 200, Lanham, Maryland 20706
www.rowmanlittlefield.com

Estover Road
Plymouth PL6 7PY
United Kingdom

British Library Cataloguing in Publication Information Available

Library of Congress Cataloging-in-Publication Data

Dowden, Bradley Harris.
  The metaphysics of time : a dialogue / Bradley Dowden.
     p. cm. — (New dialogues in philosophy)
  Includes bibliographical references.
  ISBN 978-0-7425-6030-7 (cloth : alk. paper) — ISBN 978-0-7425-6031-4 (pbk. :
alk. paper) — ISBN 978-1-4422-0059-3 (electronic)
  1. Time. I. Title.
  BD638.D69 2009
  115—dc22   2009021319

Printed in the United States of America

Contents

Acknowledgments

For their helpful suggestions, I would like to thank my students Erin Carver, Sabrina Cargill-Greer, John Daulton, and especially Matt Evpak; my colleagues Norman Swartz (Simon Fraser University) and Robert Wolf (California Polytechnic State University San Luis Opispo); two anonymous reviewers; my son, Joshua Dowden; and my daughter, Justine Dowden.

If you have an idea for improving the next edition, I'd be happy to hear from you at dowden@csus.edu. This book is fiction, except for the parts that aren't (yes, I know).

It is dedicated to my wife, Hellan Roth Dowden.

1

Introduction

SETTING: John and Naomi are riding the city bus home from campus on Thursday afternoon. John has not seen Naomi since last spring semester when she was the teaching assistant for the "Physics for Poets" course he was taking. Naomi is a physics grad student. After their chance meeting on the city bus, their conversation has drifted into swapping their favorite quotations.

Naomi: "Take your stinking paws off me, you damned, dirty ape!"

John: Excuse me?

Naomi: That's the quote.

John: Oh, I thought for a second . . .

Naomi: It's from the film *Planet of the Apes*.

John: Hmm . . . I've been on that planet. It's the one with the fur-lined teacup.

Naomi: Fur-lined teacup?

John: Yeah, you surprised me with that sci-fi quote, so I decided to do the same to you with my comment about surreal art. The fur-lined teacup is famous, like Salvador Dali's floppy clocks.

Naomi: I don't know about your teacup, but I saw a few of his clocks oozing over rocks and tables and looking like they were as soft as blankets. They made me nervous.

John: Because they're so illogical?

Naomi: Yeah. Hey, what were you listening to when I got on the bus?

John: An old song by the Chambers Brothers: "Time Has Come Today." I picked it because today was the first day of my metaphysics of time seminar.

Naomi: I've never heard the song.

John: A classic, and really long. Over ten minutes. When I first heard it I didn't realize it was longer than any other song. It's like time slowed down.

Naomi: A sign of a good song.

John: So, why're you on the bus today? Just out of a class?

Naomi: Yeah, and now I'm on my way home so I can stop drinking coffee. I've had my nose in a physics journal all day working on my research.

John: What's your research about?

Naomi: Star suicides.

John: Suicides?

Naomi: Well, star explosions. Supernovas and hypernovas.

John: So many physicists seem to be interested in things that blow up.

Naomi: Absolutely! Extreme situations tell us about the underlying physics. Explosions can be useful. Maybe a well-placed explosion on Mars could move it into an orbit closer to Earth where it would become habitable for us—not so cold. Imagine being able to choose which planet you want to live on. When this world has you down, try looking at it from another one.

John: How soon will you be seeking seed capital to develop vacation homes there?

Naomi: Ah, a philosopher who thinks like a land developer.

John: Actually it's not the land but the getting there. I'm more interested in how space travel to Mars or somewhere could affect time. The chance to study time travel is a big part of the reason why I signed up for this semester's seminar on the metaphysics of time.

Naomi: Sounds like an interesting course. When I was an undergraduate, I almost minored in philosophy while majoring in physics, but I never took a metaphysics course.

John: We've got some common interests.

Naomi: Yeah, I'm sure we do. So, what happened today in your course?

John: We started with a survey of the field. We meet once a week, and everyone has to do an oral presentation at some class meeting. I signed up to talk about time travel. The professor—his name is Henryk Mehlberg—asked us to introduce ourselves, then he talked about the definition of time. Here, look at the list of topics.

Naomi: You started out with a definition of time?

John: Well, not so much a definition as a discussion of what a definition would be like. How would *you* define time?

Naomi: I'll get back to you later on that.

John: [laughs] Then you'll tell me time is what keeps everything from happening all at once. Seriously.

Naomi: Well, let me think. Time isn't the sort of thing that can cause anything. It doesn't really cause aging, or alarm clocks to ring. OK, time is what the time variable "t" is referring to in the basic laws of physics. Time gets an implicit or indirect definition this way, don't you think?

John: Yes, but there's a problem. Think about the form of the definition. It's in the style of a definition that says time is what people are referring to when they use the word "time." It's not that these remarks are false, but that they're not especially helpful. Having a nice, brief definition of time isn't an important goal though. It's OK to start out with a definition when you're trying to figure out about time, but only as long as you haven't decided in advance to stick with the definition. The important goal is to figure out the important characteristics of time and resolve inconsistencies and controversies. That's where your laws of physics come in, but they play only a part in the grand scheme of understanding time.

Naomi: Are you saying time can't be properly defined?

John: No, but when we define a term we state its meaning. Metaphysicians who study time aren't really interested in stating the meaning of the word, at least initially. They want some deeper insight, like when some chemist had the insight that water is identical to H_2O. This was an insight about what water is, not an insight about what the word "water" means. The identity of water and H_2O successfully related the familiar to the unfamiliar in a deep and helpful way. Only later did this insight lead scientists to change their meaning of the word "water." We need something like this for time. We need a theory that provides insights about how to answer the important questions that metaphysicians want answered, and then later we can use this theory to change the meaning of the word "time" if it needs changing. Besides, even if metaphysicians were to agree that time had a certain property, they'd be likely to disagree on whether the property was contingent or necessary.

Naomi: OK, I guess holding off on a definition is a good strategy.

John: In the meantime, our class will go on to other topics. Next week is fate and knowing the future. Like, if God knows our future actions, will this deny our free will to affect the future?

Naomi: I'd like to know ahead of time whether I'm going to die on my next airplane flight.

John: Why?

Naomi: So I could cancel my reservation if there's death ahead.

John: You can't do that. If there's death ahead, then when you cancel and prevent your death, you produce a situation where there's death ahead and no death ahead. See?

Naomi: Yes, unfortunately.

John: Mehlberg says when we discuss fate he'll be showing us Aristotle's argument designed to prove we don't have free will to affect tomorrow's sea battle. We're going to have to write a two-page response.

Naomi: I hope you can figure out a way around the argument so you can save free will. There on your list in topic three—what does he mean by "mind and conventionality"? Like whether time is all in the mind?

John: That's part of it. Metaphysicians disagree over the connection between mind and time.

Naomi: I once read an anthropologist's true story about a man who lived in the countryside in Afghanistan fifty years ago. He'd agreed to meet his brother in the capital city of Kabul, but when he got there he couldn't locate him despite some serious searching. Eventually someone helped him figure out the problem. He and his brother hadn't agreed on the year they'd meet.

John: You're kidding!

Naomi: No, I swear. It happened. And what you'd be surprised about is that so many people in other cultures don't get the humor. They sympathize with the man about the miscommunication and can easily imagine it happening. Different tempos for different folks, I guess. So are you studying this sort of tempo thing?

John: Not really. That's more social psychology than philosophy, but it's interesting how different societies approach time differently. Sociology and psychology study questions such as "What kinds of people are most likely to spend their time with friends made at work rather than friends made elsewhere?" and "What's the accepted range of punctuality for a given social situation?" Like, it's worse to be very late to a dinner party than a regular party.

Naomi: Yeah, you need a better excuse.

John: The city of Cincinnati, Ohio, resisted the pressure of the railroads to have everyone within a time zone synchronize their clocks. For six years in the late nineteenth century, all Cincinnati, Ohio, clocks were synchronized to be twenty-two minutes faster than standard railroad time. Anyway, metaphysicians don't study these questions.

Naomi: I suppose they also don't study whether time exists after our death, right?

John: That's right. Instead we're going to study questions like whether time was invented or discovered. Our civilization invented football; did it also invent time?

Naomi: It's obvious to me we discovered it.

John: I think it was invented. See, there's disagreement right here on the bus, and philosophers are especially interested in areas where experts disagree. Not because disagreement is the goal, but because it shows where we need to think harder. Mehlberg said that we'll also be discussing how we tell whether an event last year lasts exactly as long as another event today. It's like asking whether last year's Christmas tree is as tall as this year's tree.

Naomi: I don't know whether the analogy is so good. It's not like we can pick up this year's event and lay it alongside last year's to see if they have the same duration.

John: Yeah, that's it. There's a big disagreement about how to make the comparison and whether the answer is conventional or instead forced on us by the way nature is. After this comes time travel, my favorite topic, and then we'll talk about whether time has an infinite past.

Naomi: I notice the experts keep changing their minds about what the universe's past is like.

John: Mehlberg says he wants to investigate what happened before the big bang.

Naomi: Good luck on that one.

John: You don't seem very optimistic.

Naomi: There won't be many traces left from before the big bang. Everybody's pre–big bang diaries are going to have burnt edges.

John: We'll be taking on an even more difficult question: Why is there something rather than nothing?

Naomi: Something rather than nothing? Do you mean why anything exists now?

John: No, the answer to that would be because of what existed yesterday. Our issue is about why there hasn't always been a completely empty universe.

Naomi: Oh. Who could know?

John: Some metaphysicians think they know.

Naomi: Hmm. Does your course have a central issue or most important topic?

John: I can't tell yet if there's a central issue. If I had to guess, I'd say it's how to answer the question, "What is time, and why does it have the properties it does?" Mehlberg believes the second half of this question is too difficult to answer, but he thinks we have a better chance with the first half.

Naomi: A physicist would say time is a one-dimensional continuum of instants.

John: I'm not sure what that means exactly, but Mehlberg mentioned this, too, and said it's some kind of smooth string of point instants, and that we'll be learning more about it. The key idea is that between any two points there's another point. His complaint was that a definition like this doesn't tell us enough of what time is, especially for metaphysical purposes. We need to resolve many other issues. One big one is whether time exists when nothing changes. No clocks ticking. No events at all. Everything is still. He says that if we don't have a position on this issue, then we don't know what time is.

Naomi: I think change is crucial to time. I mean, without it, how could there be time? Nothing would happen.

John: I don't agree. Suppose all changes on the bus stopped—no motion, no ticks of clocks, no new thoughts. Time would still go on in Russia, wouldn't it? And if it still exists in Russia, then it exists everywhere, even if we wouldn't know it does.

Naomi: Are you saying change isn't so important? It's not essential to time?

John: Yes. Well, it's important for *detecting* time. There'd be no working clocks, but without change there'd still be time itself. It would exist without our knowing it. What I'm saying is that we need to distinguish what we could know from what is.

Naomi: I'm not convinced, but I generally understand what you'll be talking about.

John: OK. Our next topic is McTaggart's argument about the unreality of time. He believes the standard approach to time, something he calls the B-theory of time, can't adequately explain time's connection with change, and that any approach that does get the connection right—his A-theory— is going to be inconsistent. Mehlberg says this issue is related to the issue of whether tensed or tenseless facts are metaphysically basic.

Naomi: You'd have to say a lot more to explain all this to me.

John: I know. I don't understand much about it either, but I will soon. Then we'll be exploring the debates about whether the past and future are real.

Naomi: Real for everybody?

John: Yes, not like when we say something is real for you and not real for me; that's a misuse of the word "real." So think about the past. It contains Socrates, don't you think?

Naomi: I've never thought about it. He's more real than Santa Claus because he's not imaginary. But he's not real as I am because he's dead. Uh, but I don't want to say he's half real and half unreal.

John: To decide whether Socrates is real, we'll need to decide whether the past is real. There are three metaphysical theories about all this. One camp of metaphysicians, the presentists, say that only the present exists. A second camp says the present and past exist, but the future doesn't. A third says the present, past, and future all exist.

Naomi: Why would anybody be a presentist?

John: Because present events are so vivid compared with past and future events. Presentists aren't merely saying the past isn't part of *present* reality; they're saying the past isn't part of *any* reality. On the other hand, the second theory, the growing-universe theory, says the present and the growing past are both real because they're both determinate, but future events are merely possible. I like this theory. The American philosopher William James said the future is so unreal that even God can't anticipate it, but I don't place that much restriction on what God can know. Anyway, there's a third alternative. The eternalist theory says that even future events are real. Actual future events, not all the possible ones. After we tackle that dispute, we talk about which objects are ontologically most basic, three-dimensional objects in space or four-dimensional events, where the fourth dimension is considered to be time.

Naomi: I wonder how you'd choose between the two, though I know in physics everybody assumes events are more basic than objects.

John: I know we'll be reading about Heraclitus to help resolve the issue. He argued in the sixth century BCE that you can't step into the same river a second time because the water you stepped into the first time is gone. Mehlberg wants us to think about whether this puzzle is best solved by assuming the basic objects are three-dimensional or four-dimensional. Puzzle solving is supposed to be a sign you're on the right path in metaphysics. The underlying problem here is how to understand change. How can something change from one time to the next? If *it*, the thing before the change, really changes, then it can't exist after the change, can it? Instead, it ceased to exist, and something else exists after the change. On the other hand, if the thing before the change is the same as afterward, then the thing didn't really change, did it? This is the big puzzle we need to solve.

Naomi: An interesting puzzle.

John: Yeah, I think so, too. Then we're going to analyze the metaphor of time's flow. Some philosophers say it's a myth. They mean time doesn't flow; it just exists. If there's any flow to time, the flow is something human consciousness grafts onto time itself. But I'm more inclined to believe the flow is real and objective and has something to do with becoming.

Naomi: Becoming?

John: With becoming real, with becoming determinate or unfuzzy.

Naomi: Is this connected with time's arrow?

John: Maybe the direction of the flow determines the direction of the arrow. Some people don't accept this, though. Anyway, the arrow is a big topic in the course. The arrow shows up when processes go one way and not the other. Like, you can scramble an egg, but you can't unscramble it. The direction that all processes normally go is the arrow of time. This arrow doesn't tell us whether time flows. There's a dispute about whether there is or isn't some physical property that gives time its arrow. And suppose all processes started going backward and the arrow reversed. Mehlberg asked us to decide whether schools would teach students the names of future presidents.

Naomi: That will be fun to think about.

John: Our list ends with Zeno. He created the ancient Greek paradox about Achilles chasing the tortoise that's crawling away from him. Zeno said Achilles' first goal is to run to where he sees the tortoise to be.

Naomi: Wait, can't we say his first goal is to run to where he expects to intercept the tortoise?

John: Yes, but that's not first according to Zeno's analysis, and he's doing the analysis.

Naomi: OK.

John: So by the time Achilles gets to where the tortoise started, he notices the tortoise has crawled to a new place. His second goal is to run to there. Yet by the time he gets there the tortoise has moved on . . . and so on. The gap between Achilles and the tortoise steadily decreases, but there's no end to the number of places he has to reach, so he can't ever catch the tortoise.

Naomi: A runner can always catch a tortoise.

John: Yes, Zeno knew it, too. Zeno's point is that there are good reasons to think that Achilles does and doesn't catch the tortoise, so our whole understanding of motion and change is screwed up.

Naomi: Oh.

John: Here's a related paradox. Switch a lamp on for a half minute, then switch it off for a quarter minute, then on for an eighth minute, off for a sixteenth, and so on. At the end of a minute there won't be any more switching. Will the lamp be on or off at the end of a minute?

Naomi: I guess it'll have to be one or the other, but I see why it can't be either because there's no last flip of the switch. I'll have to think about this.

John: Philosophers love to talk about paradoxes. The word comes from the Greek words *para* and *doxos*, which literally mean *beyond belief*.

Naomi: An apt name.

John: Mehlberg says metaphysicians should think of themselves as cooks who have to fix up a bad recipe that calls for a mix of commonly accepted beliefs, intuitions, scientific results, and metaphysical theory.

Naomi: You go to a very sophisticated culinary school.

John: I guess so. Oh, I get off at the next stop.

Naomi: Do you take this bus every day?

John: It's my first day. I stopped driving my car when I learned the grinding sound was its cry for a $1,300 transmission repair. Eventually I'll save up enough to fix it.

Naomi: Sorry about your car. I'm on this bus on Thursdays because it comes right after my section of the Physics for Poets class.

John: So maybe I'll see you next Thursday. Bye!

Naomi: See ya.

DISCUSSION QUESTIONS

1. What is the difference between what the word "time" means and what it refers to? Which of the two do you know the most about? Why is that?
2. What is the difference between the philosophy of time and the social psychology of time? Give an original example of a time question that concerns social psychology.
3. Are there any conceptual difficulties involved in a time machine's being invented that lets you (a) bring back the future cure for cancer, if there is one, and (b) meet your former self?
4. Does Socrates exist according to presentism, the growing universe theory, and also eternalism? Do all three theories allow change to exist? Why?
5. If the arrow of time reversed, would you know the outcome of a roll of the dice before the roll, and could absolutely every one of the laws of physics be the same as they are now?
6. Does Achilles complete an infinite number of tasks by the time he catches up to the tortoise? What's your definition of "task"?
7. What would our experience be like if the order of all events in time were analogous to the order of points (a) in a circle, or (b) in a branching letter Y?

FURTHER READING

Callender, Craig, and Ralph Edney. *Introducing Time.* Cambridge: Totem Books/ Icon Books Ltd., 2001.
    Covers most of the topics in this dialogue while emphasizing the relevant scientific theories. Each page is two-thirds graphics and one-third text.
Hawking, Stephen. *A Brief History of Time: The Updated and Expanded Tenth Anniversary Edition.* New York: Bantam Books, 1998.
    In chapter 2, this leading scientist informally explains the science behind our understanding of time, including Einstein's idea that an event's duration is rela-

tive to the reference frame chosen for measuring the event. "Reference frame" is defined in the glossary at the end of this dialogue.

Le Poidevin, Robin. *Travels in Four Dimensions: The Enigmas of Space and Time.* Oxford: Oxford University Press, 2003.

A philosophical introduction to conceptual questions involving space and time. There is a de-emphasis on the relevant scientific theories and an emphasis on elementary introductions to the relationship of time to change, the implications that different structures for time have for our understanding of causation, difficulties with Zeno's Paradoxes, whether time passes, the nature of the present, and why time has an arrow.

Levine, Robert. *A Geography of Time.* New York: Basic Books, 1997.

An informal introduction to the psychology and sociology of time. Page 14 contains the story about meeting the brother in Kabul.

Prior, Arthur N. "Some Free Thinking about Time." In *Logic and Reality: Essays on the Legacy of Arthur Prior,* edited by Jack Copeland, 47–51. Oxford: Clarendon Press, 1996.

Challenges Einstein's claim that time is relative to reference frame.

Smart, J. J. C. "Time." In *The Encyclopedia of Philosophy* 8, edited by Paul Edwards, 126–34. New York: Macmillan Publishing Co., Inc., and The Free Press, 1967.

A survey of philosophical ideas about time. Its section titles are "St. Augustine's Puzzles," "The Myth of Passage," "Spacetime," "Absolute and Relational Theories," "Time and the Continuum," "The Direction of Time," and "Time and Free Will: The Sea Fight Tomorrow."

2

Fatalism, Free Will, and Foreknowledge

Naomi: Hey, John. Is this seat open?

John: Sure. It's got your name on it.

Naomi: Were you just humming ABBA?

John: Yeah, I didn't realize anybody could hear me.

Naomi: ABBA is good music to dance to.

John: Or just good music for a hot day. I'm glad the bus is air-conditioned. Were you melting into the asphalt waiting out there?

Naomi: No, I timed it fairly well. Just a few minutes.

John: Hey, how was your physics class?

Naomi: Fine, nothing special, though one student asked me if my goal in life is to collect photons for a living. I'm looking forward to getting home and reading a book I started. It's a biography of John Wheeler. He invented the term "black hole" for any volume of space with matter so densely packed inside that its gravitational force prevents light from escaping. Wheeler was very interested in time travel, too. Your big topic for your class presentation, isn't it?

John: [groans]. Another student also chose the topic, and I lost the coin flip, so I had to choose again. I picked Zeno and his paradoxes.

Naomi: That'll be interesting, too. So how'd your class go today?

John: Well, we talked about fate.

Naomi: Uh-oh, the mummy of fate is getting up and opening the lid on its coffin.

John: It's not dead, but it's a very old topic and not as popular as in earlier centuries. It's worth thinking seriously about. It's not like there's been a refutation of fatalism. You see gravestones engraved with "A meeting with destiny." That's a fatalist slogan.

Naomi: Are you using the word "fatalism" as some sort of technical term in philosophy?

John: Yes, but it's ambiguous. It's the doctrine that fate exists, but the agreement ends here. Some say fatalism applies only to a few important kinds of events—who we marry, our career, our death, whether we go to heaven. It's like the fates are supernatural puppet masters pulling our strings to make those special events happen. This kind of fatalism doesn't rule out free will because we still have the ability to perform other free actions such as choosing what to have for lunch. An extreme form of fatalism implies that *all* our actions are fated; we can't do other than we do. No free will. It's like our actions are sitting out there in the future waiting to happen, and everything that happens is necessary. Philosophers are more interested in this extreme form, and often they write off the mild form as superstition.

Naomi: The extreme form isn't the same as the doctrine that we can't change the future, is it?

John: No, regardless of whether fatalism is true, no action can change the future, I'd say. Suppose a certain assassination occurs in the future. If we tried to change this by acting to prevent the assassination, we'd have to fail because the future can't both have and not have the assassination.

Naomi: I'll bet fatalists are fatalists because they make the mistake of thinking that since we can't change the future we can't affect it. So they say *que sera, sera*.

John: Not quite. They say whatever will be *has* to be.

Naomi: Oh, right. Is this extreme form the same as determinism?

John: They're similar but not the same. The future being fixed is what they have in common. Fatalists say it's fixed by fate or by some other supernatural force, and the determinists say it's fixed by any complete state of the universe at some time in the past plus the laws of nature that are operating. The laws are important because it's the laws that make causes have the effects they have. You could have fatalism in a world that had no laws of nature and so no determinism. But since both theories agree that the future is fixed somehow, they're both a challenge to free will.

Naomi: When I was growing up, my mother and I sometimes talked about this at the dinner table. Here's an old story she told. I memorized it. The master's servant went to the marketplace in Baghdad, and there he was jostled by Death. Startled, the servant returned to his master's house in fear and reported Death had made a threatening gesture in the marketplace. The master agreed to help the servant and loaned him a horse so he could flee to safety in Samarra, a half-day's ride away. Afterward, the master himself went to the marketplace in Baghdad, saw Death in the crowd, and asked why a threatening gesture had been made to his servant. "It wasn't a threatening gesture," Death replied, "only a reflex of surprise at seeing him in Baghdad, since I have an appointment with him tonight in Samarra."

John: This is about fate, isn't it?

Naomi: Yes. My mother believes God has preordained whether I'm going to heaven. He's already decided. I complained to her that whether I go to heaven ought to depend on what I choose to do along the way. If I choose to murder someone tomorrow, then God should not have decided in advance that I'm going to heaven. She responded that God has decided in advance that I won't murder anybody tomorrow. He has divine foreknowledge of his creation, she said, and if I were to do anything other than what he planned, then he wouldn't have known it in advance. Who am I to make a liar out of God?

John: Does your mother believe in free will?

Naomi: Not really.

John: Your mother's reasoning worries me. I worry that God is outside of time but with knowledge of everything that will happen, so this foreknowledge prevents me from having free will.

Naomi: I grew up thinking foreknowledge could be a problem for free will, but I don't now. Think of it this way. God's foreknowledge of my choosing to buy a cell phone depends on my choosing it. It's not that my choosing it depends on his knowledge. Let's get the dependency right.

John: I'm not so sure it doesn't go both ways, so free will is still impossible.

Naomi: OK, let me make the point a different way. I know my housemate Melissa bought a cell phone yesterday, but my knowing this didn't force her to buy it. Yes, it has to be true that she bought it, or else I don't really have the knowledge, but my knowledge doesn't force her hand. Knowledge of the past doesn't cause anything in the past, but the knowledge depends on what happened in the past. Now, turn this idea around and apply it to knowledge of the future. Knowledge of the future doesn't cause anything in the future, but knowledge of the future does depend on what happens in the future.

John: I wish I could be so confident that the idea can be turned around this way.

Naomi: I don't have a problem with foreknowledge and free will, but I do have other problems with what we've been talking about.

John: Other problems?

Naomi: Well, besides my needing a justification for assuming God exists, I have a real problem with your idea that God can be outside of time. It doesn't make sense to me.

John: I got the idea from Augustine. God's being outside of time means he has the bird's-eye view, so to speak. God is aware of all activity at all places and times. He immediately intuits all states of the universe at all times, so he doesn't need to make any inferences to gain knowledge.

Naomi: You can say all this, but I don't think you can explain it. I can say there are round squares, but I can't really explain what a round square is, can I?

John: I don't think round squares are a good analogy. I have a very clear vision of what I mean for God to be outside of time. It's like having a grand perspective on everything that happens, but having it at all times, not just at one time.

Naomi: I don't understand your vision. I'm not saying it violates logic or anything like that. It just doesn't make any more sense than saying God might bring me back to life a thousand years from now. In a thousand years there won't be any me to bring back. My old molecules will be all over the place, even inside other people. I suppose you're going to say I'm my soul and not my molecules, right?

John: Well, souls make sense, but let's get back to time. Is my vision of God being outside of time really so hard to understand? I'll admit that there are all sorts of ways people understand what God is. Thomas Aquinas and René Descartes believed God had to intervene at all times to keep the universe going. They didn't think the material world could sustain itself from one moment to the next without God's help. They thought God continually acts to preserve every material object at each successive instant; and at each instant he's using the same kind of power that it takes to create that object. This is all a little too God-intoxicated for me. I prefer a more hands-off kind of God, who watches over the material world from outside of time, but who might intervene occasionally to create a miracle.

Naomi: I still don't understand your vision of God.

John: I think you're looking for a convincing argument or justification. I'll admit that not everything I believe is based on nice, clear-cut reasons in the way you want. Usually I want those reasons too, but for a person of faith, belief in his God is usually stronger than belief in any scientific hypothesis, or in any epistemological desire for a scientific justification of their remark about God, or in the importance of satisfying any philosopher's demand for clarification.

Naomi: Well, I now feel we understand each other.

John: Wonderful! That's progress.

Naomi: You mentioned determinism earlier. I think the past influences the future but doesn't determine it, and I've got a proof that determinism is incorrect.

John: You do?

Naomi: It comes from quantum theory, our best theory of molecules, atoms, and the subatomic realm. It's our only theory of why light goes through glass and not concrete, and why copper conducts electricity better than bamboo does. Rather than predicting with certainty what we'll observe, as in the deterministic theories of relativity and Newton's mechanics, quantum theory predicts all possible results of any particular observation of nature, but it assigns probabilities to each of these results. For example, if we observe a possibly radioactive atom, quantum theory tells us the probability that it'll decay during the observation; but the instant at which it actually does decay, if it does, occurs by chance because there's no detectable difference between a radioactive atom that is about to decay and one that's still far from splitting. I mean there's no difference at all, not just that we don't have good enough instruments to detect the difference. So the probabilities are a sign that nature herself is indeterminate, unfixed, fuzzy. According to this standard interpretation of quantum theory, the heart of nature is statistical. For many individual events, there are no determining causes, just influences. Einstein was notorious for complaining that quantum theory essentially has God playing dice with the universe, and that some day a more deterministic theory would replace quantum theory. However, today's physicists are convinced that quantum uncertainty is intrinsic to nature. To put it bluntly, quantum theory has refuted the old ideas of causality and determinism.

John: I'll bet you were determined to say that because of all the sugar in your snack.

Naomi: Right.

John: OK, the way you interpret quantum theory I can see how the present will influence the future but not determine it. I just think that God knows the future even if quantum uncertainty keeps *us* from knowing it.

Naomi: He knows the outcomes of random events?

John: Yes.

Naomi: How?

John: I don't know how.

Naomi: If you don't know *how* it happens, then I don't think you can be sure you know *that* it happens.

John: You're demanding a lot from us mere mortals.

Naomi: I think you even need to worry that, if you get to heaven, God will reveal to you that everything happens for no real reason.

John: You've got to be joking!

Naomi: I was.

John: OK then, here's a serious problem for you. In class today we studied Aristotle's argument about why human action is pointless. This is his argument about a sea battle tomorrow. Suppose you have no idea one way or the other whether there will be a sea battle tomorrow. Consider both possibilities. First, suppose it's true now that there will be a sea battle tomorrow. If it's true today, then it's got to be true tomorrow. But if so, then it's not possible for me to succeed in preventing a sea battle tomorrow because it's not possible to make the truth be false. So I don't have freedom to prevent the battle. Now let's go back and consider the other possibility. Suppose instead that it's false now that there will be a sea battle tomorrow. If it's false today, then it must be false tomorrow. But if so, then it's not possible for me to cause a sea battle tomorrow because I surely can't make what's false also be true. So I don't have the freedom to cause the battle. Now, consider what we've just shown about both those possibilities. Either way I don't have the freedom that free will requires, the freedom to affect the course of future events. This argument can be generalized beyond sea battles to any other activity, so it implies the uselessness of human action and is consistent fatalism about all future events and actions.

Naomi: Do you think this is a strong argument?

John: It worries me, but I can't swallow the conclusion. To save freedom, I think we should go back and give up on the idea that a statement made now about the future is either true or false. Instead, we should say that a statement about the future is in a middle ground of "undetermined." It will become true or false later, but right now it's neither because there's no evidence available to anybody to judge it true or false.

Naomi: There's got to be a better way to save freedom than by rejecting truth-values for statements about the future. If you reject the truth-values, you'll have all sorts of trouble. Think about it. Your motivation for rejecting truth-values was that the evidence for them is lacking, but lack of evidence isn't evidence of lack. For example, there's no evidence available to anybody to judge whether Napoleon rode a horse a week before his thirteenth birthday, but do you have any problem with saying the judgment is true or false?

John: No, but that's different.

Naomi: But why treat lack of evidence about the past differently from lack of evidence about the future? Look, wouldn't you say God knows the truth-values of statements about the future?

John: Yes, but those statements don't have truth-values for us.

Naomi: So in a sense those statements do and don't have truth-values. I'd say this is some of that trouble I was just talking about. Here's more trouble. Think about what a valid argument is. In the symbolic logic course I took as an undergraduate, we evaluated arguments based on their logical form, on what the instructor called "truth's logical liaisons." For example, here's an argument with two premises and then a conclusion:
  There will be a sea battle tomorrow.
  If there will be, then we should wake up the admiral.
  Therefore, we should wake up the admiral.
Forget about free will for a second, and just assess this argument for whether it's deductively valid. Isn't it valid?

John: Yes, it's an example of *modus ponens*.

Naomi: But look what you're committed to by giving that answer. You want the argument to be valid, but it won't be valid if you deny truth-values to the premises because a valid argument is an argument in which the conclusion gets the truth-value of "true" whenever the premises do. By denying truth-values to the sentences in this argument, you're giving up on using our system of logic to analyze it. I'd say that in order to preserve logic, you should stick with the idea that predictions are true or false at the time they're uttered. Give up on Aristotle. His way out is too radical.

John: How about just changing logic? You know, expand on validity and say the argument is valid because if the sentences *were* to have truth-values and be true, then the conclusion would be true. Would that work? I thought I heard that sometimes logicians talk about changing the rules of logic.

Naomi: Yes, they do, but you should change logic as a last resort. Otherwise, it's like burning your house down and moving to a new one when you can't immediately figure any other way to get rid of the bugs in your kitchen.

John: Is logic our servant, or are we the servants of logic?

Naomi: There's a less radical way to save free will because there's another problem with the sea battle argument.

John: What's that?

Naomi: You said if it's true now that there'll be a battle, then I must not be free to prevent it. That reasoning doesn't work. If it's true now that there will be a sea battle, then I agree that there will be. But this doesn't rule out my freedom to prevent it.

John: I don't understand.

Naomi: Think of it like this. If it's true now that there will be a sea battle, then it's not true that there won't be. OK?

John: So far so good.

Naomi: But from this, you can't legitimately infer that it's not *possible* that there won't be. That's the mistake. And if it's possible that there won't be, then I can act to affect which possibility becomes an actuality, so how can you go on and infer that my hands are tied and that I don't have free will in the matter? In other words, if it's true that there will be a battle, then I'm free to prevent it even though in fact I won't. See what I mean?

John: No. It's already true ahead of time what will happen; you can't make it not happen. Free will is the ability to do otherwise, but you aren't free to do otherwise, so you don't have free will.

Naomi: OK, I'll say it another way. If it's true, it's not false. But if it's true, it's a mistake to say it's not possible that it's false.

John: I'm not sure what to say. Wait! Here's a better way to look at it. Suppose there will be a sea battle tomorrow. Then think about all the possible future situations from now. Here we are at time one with the universe having progressed to the point where it's true there will be a sea battle tomorrow. That means that all possible futures containing the universe as it is at time one also contain a sea battle tomorrow. Right?

Naomi: Yes.

John: Now, if you say you can prevent the sea battle tomorrow, you're saying there's some possible world in which there isn't a sea battle. But that's contradictory because you just agreed a second ago that all possible futures contain a sea battle. So Aristotle's argument succeeds. I mean it would succeed unless we deny truth-values to predictions.

Naomi: I'm not convinced that your reasoning shows Aristotle succeeds because I don't like how you're treating the concept of "ability to do otherwise." Also, you're recommending what my philosophy of language instructor called a "tensed view of semantics" where statements about the future don't now have truth-values. That's overkill. To find a way out of Aristotle's problem, we just need not to make errors in reasoning about possibilities and to get right what free will really is. Free will is really something that takes time to reveal itself.

John: Yes, you need some evidence over a period of time that shows the ability to do otherwise in the same situation. I'm talking about situations where you have some choice and where there isn't external compulsion such as being hypnotized or having a gun held to your head. That's free will.

Naomi: Not quite.

John: What now?

Naomi: I don't think free will requires the ability to do otherwise in the same situation.

John: You don't?

Naomi: It just requires the ability to do otherwise in very similar situations so that you detect plasticity of action. Here's what I mean. Imagine some situation where you're sure Sam is acting freely. Maybe you give him a choice of soda drinks and watch that he doesn't act as mechanically as a soda machine. He doesn't, so you correctly draw the conclusion that he has free will. But wait. What you and Sam didn't know was that God earlier sent down an enforcer-angel to watch over Sam. The angel has been hiding in Sam's head and was going to prevent certain choices by Sam if he showed an intention to choose other than the way the angel wanted. But by luck Sam happened to have behaved the way the angel wanted. Wouldn't you say that in this angel situation Sam could still be acting with free will?

John: Yes, I guess so, though it's an odd situation.

Naomi: But Sam couldn't have done otherwise! Doesn't this story about Sam and the angel show that having the ability to do otherwise in the same situation isn't necessary for free will?

John: Yes, I guess so. But if you're right, then we need to figure out what exactly is necessary for free will—other than not having external forces making our choice for us. It's got to be something else.

Naomi: I've got a suggestion. First off, it's really necessary that you be affected by external forces. Suppose you were given a choice of Coke or Pepsi but were then told, and you understood and believed, that if you don't choose Pepsi, then you and your family will be murdered. Now in this choice situation if you went ahead and chose Coke, I'd be worried that you were just a machine and had no free will in the situation. If you couldn't be affected by threats, there'd be no "plasticity" in your actions, and so no free will.

John: OK, that seems right. You've got to be capable of being influenced, or you don't have free will.

Naomi: Yes, and second of all you have to have the ability to do otherwise in very slightly different situations.

John: Like what sort of situations?

Naomi: Let's go back to the sea battle scenario. Suppose at some time, call it time one, you're faced with a decision about whether to start a sea battle, and you decide at that time to start a sea battle. What "ability to choose otherwise" really means is that in some possible world similar to, but not exactly like this one, you make a choice to try to prevent the sea battle. In the similar world, there are no new external constraints on you, and all that's significantly different is that you change your intentions about whether to attack, and this change involves a change in your brain that is presently too subtle for anybody else to detect. No contradiction. That's real freedom.

John: Why are you saying the change is too subtle to detect?

Naomi: It's got to do with chaos theory. Think how hard it would be to roll a pair of dice a second time and be absolutely sure of getting the same total as the first roll. The outcome can be radically different with the very slightest change in initial conditions. That's a sign of chaos at work. Probably there was some slight but unmeasurable difference between the two situations at the beginning of the roll, and this is why you got a different total the second time. Now, suppose you ask me whether I want a Coke or Pepsi. I might choose Coke the first time and Pepsi the second time even though you try to set up the situation the second time so it's just like the first time. You can't be sure of setting it up again with me having the same intentions because the brain traces of my intentions aren't sufficiently detectable and controllable by you, but my slight change in intention can precipitate a big difference in my actions involving choice of soda. You can carefully monitor my previous behavior and use the evidence from my brain scans, but you can't get enough information to predict with certainty. That inability to control the situation is what I was talking about when I said you get a different outcome when the present situation is a little different in a certain way. When you ask me to make a choice the second time around, I might surprise you; that's the sign of free will. Soda machines can't surprise you.

John: You're saying free will is just chaos and surprise in situations without external compulsion?

Naomi: No, that's an exaggeration. Not just any old surprise. It's not like your intention to choose Coke leads to such a chaotic result that the surprising words "Chartreuse isn't yellow" might pop out of your mouth without your intention to say this. That wouldn't be free will at work. The surprise has to be of your own making; it has to be what you intend. The action is a surprise to other people but not to you. Of course, an outside viewer could force the same outcome by pointing a gun at your head and saying, "You better make the same choice as last time or you'll be shot." But so what? That wouldn't be a situation where you'd expect free will to show itself, would it?

John: No. But there are a lot of other theories of free will out there, and I'm still not convinced that yours is the right one. For one thing, it's compatible with determinism, isn't it?

Naomi: Yes, even though I don't believe in determinism.

John: Oh, this is my stop. I'm off.

Naomi: Off to meet your destiny, no doubt. Hey, don't forget your phone.

John: [standing up] Oh, thanks. I can't live without it. All my music is in there. See you next week.

Naomi: Bye.

DISCUSSION QUESTIONS

1. What does free will have to do with time?
2. What is fatalism, and how does it differ from determinism?

3. Do you have any good evidence that the way you'll die in the future isn't fated?
4. What part of your future can be changed?
5. Could God know in advance the outcomes of your free choices? If someone disagreed with you about this, what sort of reasons would they be likely to offer?
6. Does Aristotle's argument about a sea battle tomorrow show that the future is fixed and free will is impossible? If someone disagreed with you about this, what sort of reasons would they be likely to offer?
7. When we try to answer the question, "What does reality consist in?" should we say it consists in part of (a) free will in some nonhuman animals, (b) trillions of years, (c) truth-values, and (d) sakes (as when we say he did it for her sake)?
8. What do John and Naomi have to say about whether free will requires the ability to do otherwise? What do you think?

FURTHER READING

Barnes, Jonathan, ed. *The Complete Works of Aristotle*. Princeton, NJ: Princeton University Press, 1984.
De Interpretatione (On Interpretation), Book 9, sections 18a–19b, is the source of Aristotle's third century BCE argument that statements now about tomorrow's sea battle lack a truth-value.
Blackburn, Simon. *Truth: A Compelling Introduction to Philosophy*. Oxford: Oxford University Press, 1999.
Pages 81–119 present a compatibilist definition of free will that works like this: The subject acted freely if she could have done otherwise in the right sense. This means that she would have done otherwise if she had chosen differently *and*, under the impact of other *true and available* thoughts or considerations, she *would* have chosen differently. True and available thoughts and considerations are those that represent her situation accurately and are ones that she could reasonably be expected to have taken into account.
Boethius. *Consolation of Philosophy*. Translated by Joel C. Relihan. Indianapolis, IN: Hackett Publishing Company, Inc., 2001.
This early sixth-century Roman philosopher creates a dialogue between himself and a woman named Philosophy. In the dialogue, Boethius worries that, if God knows now what you will do later, then what you will do is already determined and so is not free. Philosophy counters by arguing that God is outside of time and so knows all your actions atemporally, and this way of knowing does not require your actions to be determined.
Butterfield, Jeremy. "Determinism and Indeterminism." In *Routledge Encyclopedia of Philosophy*, edited by Edward Craig. London: Taylor & Francis Group, 1998.
Argues that the doctrine of determinism is commonly misunderstood. For example, much of Newton's physics is indeterministic. Focuses on arriving at

determinism from an analysis of scientific theories and claims that formulations of determinism in terms of causation and predictability are unsatisfactory.

Craig, Edward. "Fatalism." In *Routledge Encyclopedia of Philosophy*, edited by Edward Craig. London: Taylor & Francis Group, 1998.

A very short discussion of the variety of meanings of the term "fatalism."

Frankfurt, Harry. "Moral Responsibility and Alternate Possibilities." *Journal of Philosophy* 66 (1969): 829–39.

The source for the scenario with the enforcer-angel that has been hiding in Sam's head and was going to prevent certain choices by Sam if he showed an intention to choose other than the way the angel wanted.

Haack, Susan. *Deviant Logic*. Cambridge: Cambridge University Press, 1974.

Chapter 4 contains a clear account of Aristotle's argument for treating the scenario of the sea battle by denying truth-values to future contingent statements.

Le Poidevin, Robin. "The Experience and Perception of Time." *Stanford Encyclopedia of Philosophy*.

Explores how metaphysical theories of time accommodate various aspects of our experience. The entry discusses the flow of time, an issue discussed in a later chapter of this dialogue.

Maugham, W. Somerset. "Appointment in Samarra." *The Complete Short Stories of W. Somerset Maugham*. London: Heineman, 1951.

Contains the story in the dialogue about Samarra and fate.

Swartz, Norman. "Foreknowledge and Free Will." *The Internet Encyclopedia of Philosophy*.

Careful examination of the apparent inconsistency between divine foreknowledge and the human exercise of free will.

3

Mind, the Metric,
and Conventionality

Naomi: Hi, John.

John: Welcome back.

Naomi: ABBA again?

John: Nope. "Thriller."

Naomi: I love the early Michael Jackson stuff. Hey, another scorcher, isn't it?

John: Yeah, whenever I was outside I felt like the ant being burned under the magnifying glass.

Naomi: At least our air-conditioning is working. How's your day otherwise?

John: Great. I'm richer now.

Naomi: Yeah? How much?

John: My supervisor at the restaurant where I work left a message on my cell phone today saying I'm getting an 8 percent raise. I'm going to start eating two meals a day instead of just one.

Naomi: Uhhh . . .

John: Just kidding about the meals. So how're you doing? Any success in understanding exploding stars?

Naomi: A little. I have a wastebasket full of crumpled ideas. My plan is to throw away all my bad ideas, and eventually this will leave the good ideas.

John: Great plan. I know you were joking, but it's sort of the way I look at philosophy. Even if I'm not able to find out exactly what is true, it can be very helpful to find out what is false.

Naomi: Yeah, that's a kind of progress. So how'd your class go today? Any insights into time?

John: We discussed whether time exists, and how it relates to mind. If you made a list of all the entities that exist objectively, why would you put *time* on the list?

Naomi: Because it exists out there.

John: Yes, but how do you know that? We discussed this a little bit in class today. I'm not sure time does exist objectively, but one of the other students in my class had a proof. Change implies time, and change definitely exists objectively. I think it was Aristotle who said time is the measure of change.

Naomi: That's the proof?

John: The first part of it. Second, and just as important, you and I and nearly everyone else can agree on the order that events occur in. Lincoln's birth first; his death later. Pay taxes first; get refund later. Who would disagree? If we're judging in the same reference frame, we can agree that all events line up on one line, in one dimension.

Naomi: OK, that's got to be the heart of it, but I think there's more to the story.

John: Like what?

Naomi: We use time in so many of the formulas of physics, and these formulas—they're actually scientific principles—are part of our best theories of the world. These theories have all sorts of practical consequences; they help us land on Mars and explain why radios work. It would be a miracle if the theories were empirically successful like this but there really were no such thing as the time we thought we were referring to with our formulas. And it's important that the formulas are about so many different kinds of phenomena. Our universe has a large number of different physical processes that bear consistent time relations, or frequency of occurrence relations, to each other. For example, the frequency of a fixed-length pendulum is a constant multiple of the half-life of a specific radioactive uranium isotope; the relationship doesn't change as time goes by, at least not much and not for a very, very long time. The two processes have a special relationship in that they "tick" along together. Another example would be that the number of times per second a magnet is pushed in and out of the center of a long coil of wire is directly proportional to the voltage induced in the wire. The existence of these sorts of stable relationships involving time makes our system of physical laws much simpler than it otherwise would be, and it makes us more confident that there is something real and objective we're referring to with the time-variable in those laws.

John: OK, so the argument is that time exists objectively because it's implied objective change and because we infer that it's the best explanation of the existence of change, and of our having so much agreement on what occurs before what, and of our success using the formulas of physics involving time to cover a wide variety of phenomena. But I have a problem with your reliance on the formulas of science. Can't your formulas be empirically successful and still be false?

Naomi: Yes, but they aren't successful by accident. There's a big difference between a useful theory like the flat earth theory and a silly theory like the earth's being a checkerboard with all the black squares missing, and this tells me the flat earth theory had some degree of truth, at least for small areas of the earth's surface. I suppose you're looking at the history of science as if it's a graveyard of falsified theories, so our current theories can be expected to die, too.

John: Yes.

Naomi: OK, suppose we say that the empirical success shows that our theories and their formulas are *approximately* true instead of definitely true.

John: Now you've got a stronger argument for why time is real.

Naomi: Thanks.

John: How do you suppose people came to believe that time is real in the first place? What suggested it to them back before the scientists invented their formulas and even before somebody built a clock? I mean, nobody would build a clock unless they believed there is something like time that exists and can be measured by a clock, right? I think the story goes like this. Although time is abstract and not directly accessible to our senses, our ancestors always had an indirect ability to sense time using some structures inside their brains. Those structures evolved in their more distant ancestors due to their living on a spinning planet with periodic cycles of daylight and darkness. Our ancestors sensed the rhythm. So people came to believe in time long before the first clock was invented, but it was not a firm belief. It got firmer when they figured out that they could actually use a clock to measure time. Clocks are just devices that count the periods of some regular periodic process.

Naomi: I like that explanation of why people came to believe time is real.

John: Thanks.

Naomi: Our internal psychological clock is inaccurate, but it's interesting how people are so good at estimating how much time has passed. When we sleep, you'd think our time sensor would be turned off, but it's not; it's just inaccurate. When we wake up, we sometimes don't know if we've been asleep for two hours or six hours, but we never have to worry that maybe it's been six months.

John: Never? My first philosophy teacher, Wendell Johnson, said "always" and "never" are two words you should always remember never to use.

Naomi: Almost never.

John: Did you hear about the unmarried woman who went to see her doctor and was told she had only six months to live? The doctor told her to go out and marry a tax attorney. "Why a tax attorney?" she asked. He answered, "Because that'll make six months seem like an eternity."

Naomi: Ha! I hope I can remember this so I can tell my housemates. . . . I wonder how bears know when to wake up from hibernation after six months.

John: Bears sense time passing, but I'm not so sure you should say they know when to wake up. Having the ability and knowing you have the ability are different. And I don't think bears have an actual conception of time; they can't think about time.

Naomi: OK, bears are dumb. Do you suppose that when bears sense time, it feels to them on the inside the way it feels to us?

John: I once would have said, "Who can know? We aren't bears." But now I'd say their sense of time has to be similar to ours because we're both mammals. The internal experience between two beings is more alike if their biology is more alike. A turtle's sense of time should differ more from our sense.

Naomi: Yeah. A snail was mugged by two turtles. When the police asked him what happened, he said, "I don't know. It all happened so fast."

John: [laughs]

Naomi: It's an interesting question in animal psychology what sort of durations an animal can react to. A friend of mine who works in an animal lab says a rat can learn to press a lever that will reward it with food after a short delay, but if the delay is over half a minute, it can't ever learn to get the reward.

John: I suppose it's a memory problem. I'm sure squirrels are better than rats at this lever experiment. They remember where they buried their nuts months ago. Do you suppose animal time comes in degrees or is it an off-or-on kind of thing?

Naomi: I've no idea, but it would be interesting to find out.

John: In class, we talked about this and eventually agreed it was an ambiguous question. To sort things out, we decided it's very helpful to draw a distinction between physical time and psychological time. Physical time is what clocks are designed to measure. It's public time. Psychological time is a being's awareness of clock time. Psychological time is its private sense of public time. A rat's psychological time is probably very different from our psychological time, but we both have to live by the same physical time.

Naomi: That's a good distinction to make. Clocks aren't affected by any mental states.

John: I think when someone says, "Do you have time to do this before lunch?" they're being ambiguous and might be referring either to psychological time or to physical time; but when poets mention time in their poems, they're almost always referring to psychological time, don't you think?

Naomi: Yes, but I'd say physical time is definitely more helpful to us than psychological time when it comes to doing science. When we define average speed along a path to be the measure of the distance traveled along the path divided by the measure of the time it took to complete the path, we don't want to use someone's psychological time.

John: I agree, but actually the psychology of time is as interesting to me as the physics of time. Unfortunately, Mehlberg says our civilization knows considerably more about physical time. Cognitive scientists don't know which neural mechanisms account for the experience that everybody has of time's flow. And we go around all day blinking now and then, but our mind somehow adjusts to this and removes all that blacked-out experience while our eyelids are closed. Nobody knows how the blacked-out time is handled. Then there's all that human agreement on the time order of events in our daily lives, but nobody knows how we represent this time order in our brains, though the cognitive scientists say the representation must have something or other to do with our neurons and how they connect up.

Naomi: I'm sure they're right. There are plenty of future Nobel Prizes to be won here.

John: We did talk today about one idea of how the mind's psychological time might be connected to physical time—through memory.

Naomi: How's that?

John: Well, memory is all about traces of the past left in the brain. We have traces of the past but not of the future, and our brains use this fact to make the connection with how processes evolve in the outer world. Early in our life, and maybe even by birth, we come to implicitly understand that event 1 happens in time before event 2 if, at the time we become aware of event 2 occurring, we already have a memory of event 1 occurring, but not the other way around. In this way we make a distinction between before and after in psychological time that tracks the way causes come before effects out in the external world.

Naomi: Yeah, that sounds right. It's got to be a big part of why we have psychological time and why our sense of the order of temporal events isn't too far off from the way physical time really goes. Memories can't be trusted all by themselves because a person can falsely remember the order that two events occur in. Overall though for our society, the causal ordering brings coherence to the remembered ordering even if we have to say a few memories are mistaken and that psychological time deviates now and then from physical time.

John: Mehlberg said the most surprising scientific discovery about psychological time is Benjamin Libet's experiments in the 1970s showing that unconscious brain events involved in initiating a free choice occur about a third of a second before we're consciously aware of choosing. Before Libet's work, it was universally agreed that a person is aware of choosing to act freely, then later the body initiates the action. If Libet's interpretation of his experiments turns out to be correct, it's going to upset a lot of thinking about free will and human dignity.

Naomi: Hey! You didn't warn me to wear a seat belt before you took that radical turn away from common sense. Think about my free choice this morning to reach for the strawberry jam instead of the grape jam. You're implying my awareness of choosing strawberry occurs *after* my brain circuits start sending signals to my arm to reach for the strawberry jam! Right? I hope researchers are double checking this.

John: The experiments have been re-run, but there's still controversy about how they should be interpreted. There's so much yet to understand about the relationship between mind and time.

Naomi: Yeah.

John: You and I have agreed that some reasons suggest that time might exist objectively, and we agree that psychological time is awareness of physical time, and that physical time is linear, and that it's useful for science, but we haven't figured out whether physical time depends on our minds. If it does, it isn't objectively real.

Naomi: All our thinking about time takes place in our mind and is expressed in our language, but that's no reason to believe that either mind or language has anything to do with making time what it is any more than they have to do with making copper wires able to conduct electricity.

John: The issue of the relationship between time and mind is more complicated than this.

Naomi: Uh-oh. Am I going to need a seat belt?

John: Yes.

Naomi: Are you going to say that if nothing outside of my mind is real, then physical time isn't either?

John: No, but that's true. I'll agree with you that the evidence for saying time is real and objective is all that agreement among people about the time ordering of events and all the agreement about the empirical success of scientific formulas that use a term for time. But not only do we have to worry about our scientific theories being empirically successful while not being literally true, the deeper question is whether we get this agreement because all our minds work the same way and not because we're all detecting how things go out there in the universe external to our minds. Here's the worry. Humans who aren't color-blind agree that the color of ripe strawberries is red, but it's not obvious that the red color exists objectively in the strawberries. The colors are real and not imaginary, but aren't they partly subjective? Kill everyone and strawberries are no longer red, though they continue emitting the same frequency of light waves. And if colors turn out to be partly subjective, despite all our agreement, then how do we know time won't turn out to be, too?

Naomi: Kill everyone and the strawberries are still red because they're still *capable* of producing the red experience in someone who is built like us. Red is about capability, not about who's living. But even if I were to agree with you that redness is mind-dependent, what reason is there to think time is like that?

John: In the late 1700s, Immanuel Kant argued that our mind actually structures our perceptions by making us have our experience in time.

Then by a priori reasoning about how it's possible for us to have any perceptual knowledge at all we can figure out that time exists. And by even more a priori reasoning we can figure out that time has the structure of a mathematical line. If you examine the very preconditions for a human mind having any experience at all—he was pretty vague on how you do this—you discover a deep connection between time and arithmetic, he said. It's not that times or instants obey the laws of arithmetic, but the other way around. He said that the foundation of arithmetic is in the temporal structure of our experience.

Naomi: I don't think there's an a priori connection between time and arithmetic.

John: Most things Kant said were not said very memorably, but when he says time is an a priori form of intuition or a form of conscious experience, I think he means we humans have no direct perception of time but have only the ability to experience things and events *in* time. He had the same kind of idea about space. He said our mind structures our perceptions so that we experience objects in space, and we can use a priori reasoning to figure out that space always has a Euclidean geometry. This treatment of space and time is called transcendental idealism. That's his terminology.

Naomi: I see a major problem. Euclidean and non-Euclidean geometry disagree about parallel lines. Pick a line and a point not on the line. How many parallel lines go through the point and never cross the original line?

John: One.

Naomi: Euclidean geometry says one. Non-Euclidean geometry says it's not one. Who can tell which of these two geometries is the correct theory of space? Well, Einstein could tell, but not by a priori reasoning. His theory of relativity adopted non-Euclidean geometry, and it had all kinds of success and has become our most fundamental theory of nature alongside quantum theory. So basically Kant lost out when Einstein and non-Euclidean geometry came along. That means we can't trust what Kant says about our having a priori knowledge of time. That's what I meant when I said that I see a major problem.

John: OK, you don't trust Kant's transcendental idealism.

Naomi: Do you and Kant believe time is objective?

John: Well, I don't, but Kant would say we have objective knowledge of true statements about time. The problem with saying this is that he doesn't look at objectivity as straightforwardly as you do. He uses the term oddly. Even if I were to agree with you that time is objective if anything is objective, I'm not convinced yet that it's even possible to give an objective representation of what exists, in your sense of "objective." Here's what I worry about. The goldfish is always stuck with a fishy perspective that keeps it from knowing the way things are, for example that fish and humans had a common ancestor species millions of years ago. No fish will ever know this. By analogy, maybe humans are always stuck with a peculiarly human perspective and therefore can't see reality as it is. We have these human categories forced on us just because we're human, and so objectivity always eludes us. It's very humbling to think this about the human situation, and it's also very depressing. It's like we're forever trapped in a condition of human subjectivity, and only God can escape to see things as they really are. I would like to know how God sees time, but I'm sure he couldn't tell me in a way that I could possibly understand. This position of mine is sometimes called "cognitive relativism."

Naomi: So in order to say whether time is objective we have to figure out whether we can have any objective knowledge at all?

John: Yes, and the issue has another feature. In the twentieth century, the philosopher of science Bas van Fraassen described time—wait a second, I took notes—by saying there would be no time were there no beings capable of reason, just as there would be no food were there no organisms and no teacups if there were no tea drinkers. He means there's no society-independent fact of the matter as to whether a piece of clay is a teacup, and no biology-independent fact of the matter as to whether that stuff in the supermarket is food, and no mind-independent fact of the matter as to whether the system of physical changes out there is time.

Naomi: Do you agree with van Fraassen?

John: Yes, definitely. What about you?

Naomi: I'm on the other side of the fence. Next he's going to be saying there'd be no stars if there were no stargazers. He's fallen into the swamp of anti-realism. The only point I'd concede to this van Fraassen guy is that there'd be no time if there were no changes. But he wants time to be dependent on human reasoning. I'm sure time exists with or without humans. Consider this time order: first, a stone hits the surface of a still pond; second, the waves reach the shore. This order would not reverse if beings capable of reason had never evolved to represent time. If there were a physically possible world in which all the usual orderings of events were there but no conscious beings existed to represent them as time, then it would be a mistake to say it's *merely as if* the world contained time; it would really contain it. I guess what I think is that natural science deals with observer-independent phenomena and social science deals with observer-dependent phenomena, and time can be studied by *natural* science.

John: It's hard to know what the universe is like independent of our culture and our language. It's even harder to know what the universe is like independent of conscious mind. I think consciousness is some lens that affects what we can know. We conscious beings have stones in our universe, but does the universe have stones without us and our culture?

Naomi: It's the same universe.

John: Is it? I don't think you can check up on this to tell if you're correct.

Naomi: You metaphysicians challenge everything! I recommend hanging up a sign over the door to your metaphysics seminar like the one in Dante's underworld, saying, "Abandon all hope, ye who enter here."

John: It's just challenging, not hopeless. Here's another possibility to consider. Isn't it possible that we humans don't realize that our universe as we detect it is merely an illusion created in our brain and that our brain is really floating in a vat of liquid nutrient somewhere and our brain is plugged into some supercomputer controlling our thoughts?

Naomi: That's the ultimate in paranoia. Do you have any real reason to suspect you're a brain in a vat?

John: No, but it's possible, isn't it?

Naomi: If I'm going to agree that it's possible, then I'd want there to be some imaginable way to figure out whether it's true.

John: Well, maybe we can't figure out now whether it's true, but we'd know who was right if someone started unplugging our brains from the computer. Then we'd suddenly realize that we were floating in the vat, or at least we would if the person hadn't yet unplugged our visual input system. And it could be even worse.

Naomi: Worse than that?

John: Maybe we don't even have brains, vat or no vat. Instead, we exist simply as a network of computer circuits within a world simulation program run on some video game player's computer in an advanced civilization somewhere. Maybe we're just virtual players in somebody's computer game. I'd say there's a chance that this is the human condition we find ourselves in.

Naomi: This is really far-fetched.

John: I agree completely, but that doesn't mean it's not worth thinking about.

Naomi: I agree these scenarios can't be ruled out, but all this worrying about the influence of the mind doesn't make me want to change my mind and say time isn't objectively real.

John: This whole disagreement between realism and anti-realism definitely isn't settled. But how about this as a way to understand time? We look around our world and see how convenient and useful it is to have a certain conception of time, so what's really going on here is that we *choose* the time that is used in those physics equations of yours. See the problem? That makes our time the most convenient time, not the true time.

Naomi: I think there's one true time. The convenience is a sign to me that we've sniffed out the truth. Human choice is involved in our getting precise about any of our concepts, including the concept of time, but that doesn't make everything conventional. It's convenient to agree to divide up the universe into fish and non-fish, and then say that all fish swim. You can't go and say this makes it *conventional* that all fish swim.

John: You've got a point, but here's the problem the way I see it. We make our best choice about how to describe time. Is this choice also a process of getting at the truth of the matter, or is the choice conventional, or is it a bit of both? That's the question. A choice to use the English system of units instead of the metric system is all conventional or a matter of politics and economics. One system isn't truer than the other. Is choosing a best theory of time more like choosing to use the metric system, or is it more like choosing to believe that all fish swim?

Naomi: We choose all our concepts, but some of the concepts are sort of forced on us because we'd never understand the world without them. Time is one of those concepts.

John: I wonder if there's an ambiguity here. We use the word "concept" differently when we say, "The concept of time is fundamental in physics," from when we say, "I didn't have a clear concept of time until I studied philosophy." The first is objective; the second is subjective.

Naomi: OK, when I was talking about the concept of time I didn't mean someone's subjective conception of time. The deep question here, as I see it, is with clock time and whether there's a certain arbitrariness or conventionality in choosing its metric and in choosing the standard clock.

John: What do you mean by "choosing the metric"?

Naomi: It has nothing to do with the metric system. It's about choosing the lengths of time intervals, or what we call "durations." It's about whether the duration between two events yesterday is the same as the duration between two other events now. The decision to say yes or no involves our choosing a metric. The way we do this is by placing numbers on the two durations and saying whether the numbers are equal.

John: I need more details to understand this.

Naomi: When we choose a coordinate system for our reference frame, what are we doing? We're basically choosing some event to be at the origin and to have the time zero. The coordinate system associates time numbers with all events throughout space and time. It assigns single real numbers to instantaneous events. These single numbers are the time coordinates. Like, this happened at 5 PM and that happened earlier at 1 PM. The metric comes in when we consider events that are not instantaneous and want to say how to use those time coordinates in order to figure out how long the event lasts or to figure out whether two events lasted the same amount of time. We need a rule to tell the duration, that is, the measure of the time interval, between two instantaneous events when we're given their coordinate numbers, their times of occurrence. The rule is the metric. Usually the rule is simply to subtract the smaller number from the bigger. The conventionality question arises when we ask whether we could have chosen some other rule than the subtraction rule.

John: That was a little fast. Can you give me an example?

Naomi: Sure. When the clock says it's five o'clock, how much time has passed on the clock since it last showed one o'clock? Four hours—because that's five minus one. The metric is this principle of subtraction. Now, how about the time of processes that aren't inside the clock? The idea is that clocks don't just show the time of their own processes, they show the time of nearby processes. If we attend a sports event and notice our watch changed from 1 PM to 5 PM, then we say the sports event itself has lasted four hours. Children are taught how to measure intervals of time by reading a watch this way, but we don't tell them the gory details about their using a metric, and placing a coordinate system on point events, and assigning real-numbered times to those events. The metric is really important, though, and it's why we say the correct answer is "four." Why didn't we say the duration of the sports event was five minus one divided by seven, or why didn't we take the logarithm of the difference in the two numbers? What's so special about simple subtraction?

John: I'm not sure. Maybe I'm making some unconscious choice, but it just seems natural.

Naomi: It's natural to me, too, but more complicated functions also do the job of being a metric. The job of being a metric that I'm talking about is the job of satisfying three requirements: one, the requirement that the distance between two points x and y be zero just when x *is* y; and two, the requirement that the distance from x to y be the same as from y to x; and three, the requirement that the distance from x and y plus the distance from y to z never be smaller than the distance from x to z. There's general agreement among the experts that metrics or distances should satisfy these three requirements. Distance in space is called "length," and distance in time is called "duration." The simplest way to satisfy these three requirements on a metric for time, since it's one-dimensional, is to say the duration from x to y is the absolute value of the difference in x and y. And the deep philosophical problem is whether the simplest way is the correct way or instead is merely the conventional way. Also, we need to carefully distinguish the question of whether time is conventional from the question of whether time's metric is conventional, because even if the metric were conventional, it might not be conventional that the universe is structured so that there exists a time that can have a conventional metric.

John: That's clearer. You're saying our civilization has chosen the simplest metric among other equally usable metrics. But I'm not sure I understand the point about *other* choices being OK. Are you saying that we could choose a metric that says that the difference between 1 PM and 5 PM is four hours on Tuesdays but a half-hour on Wednesdays?

Naomi: No, because a choice like that would violate one of those three assumptions about what a metric is. What I'm saying is that we could have picked a metric in which the duration between the specific pair of events on Tuesday was seventy-seven times the difference in their coordinate numbers. And that choice would impact Wednesday events.

John: So you're not saying that just any silly thing is OK here.

Naomi: Right.

John: OK, but it seems to me that if time weren't continuous but instead were atomistic, then you could find out whether intervals last the same time by counting the finite number of atoms in the intervals. The only metric allowable would be the count.

Naomi: I agree that the allowable metrics for discrete time are severely limited. But with ordinary nonatomic time, there are additional places in the story where it's been claimed that convention creeps in. We assign real numbers to instantaneous events, and we assign intervals of real numbers to noninstantaneous events. Well, why real numbers and not rational numbers? Human measuring devices can't distinguish real numbers from rational numbers, and for this reason some people argue that the choice of real numbers is conventional. But I think they're wrong.

John: So you're saying there are facts about the universe that require time to be measured with real numbers rather than rational numbers?

Naomi: Yes.

John: What are those facts?

Naomi: I can't give you a simple answer. The reasoning is very indirect. For scientific purposes, it's more elegant to have scientific laws using real numbers than rational numbers. I'd hate to have to do science without real numbers.

John: Well, if that's the reason, then I'd say the choice of real numbers over rational numbers is definitely a source of convention. Convention must be all over the place in science and mathematics because scientific principles are underdetermined by all the data from all our observations of nature. I'm thinking of a graph containing a finite number of data points, with a line through the points being the representation of the scientific principle. Isn't there always more than one way to draw a continuous line through those points?

Naomi: Yes, there are an infinite number of ways. Nearly all our supposedly true scientific generalizations are underdetermined by the available evidence, yet we realists still think they're true or false and not simply conventions. Compared to features about galaxy structure, features about time structure, such as whether it's continuous, are more indirectly connected to any data producible by a scientific experiment. However, I know many anti-realists believe scientific principles aren't giving us any truth at all but only useful tools for meeting our needs.

John: I've always been attracted to this kind of anti-realism.

Naomi: I'm an objectivist about choosing the metric and choosing real numbers over rational numbers, so you probably disagree with me about clock choices.

John: What do you mean?

Naomi: Do you know the song by Chicago called "Does Anybody Really Know What Time It Is?"

John: Sure, I can sing it: "Does anybody really know what time it is? Does anybody really care? . . ."

Naomi: Yeah, that's it. Suppose we do care and want to know what time it is. First off, we'd look at a clock, right? It's designed to tell us the time coordinates along the time dimension. When we use our clock, we normally assume that any two intervals between successive clicks will last the same amount of time. That is, we assume that the clock stays regular. But how do we know our clock really does have regular periods? Maybe it slows down a little each month. What would you say is the way to know?

John: Well, I'd compare the clock to a better clock.

Naomi: Yeah, that's it. Ultimately we'd want to compare it with the best clock, our standard clock, which ticks regularly by definition. The standard clock is a clock whose intervals between any two successive clicks are always the same. All the other clocks we use have to stay synchronized with the standard clock, or else their readings aren't accurate.

John: Now, I'm trying to figure out the relationship between the standard clock and the metric. Is it that choosing the standard clock is claiming that there's a machine that can always select a distance in time of the same length?

Naomi: You've got it! Now here's the big question. Assuming we've picked the metric, is it merely a convention that we chose one candidate for a standard clock over another candidate?

John: Sure, because it's a convention to say it can't show the wrong time. The standard clock can never be inaccurate because it defines what we mean by "accurate." Selecting the clock means we are arbitrarily declaring how long a second lasts.

Naomi: Are you saying it's conventional that a second lasts for, let's say, 622 ticks on the standard clock and not 644?

John: No, that's about what *unit* we pick for measuring time in our coordinate system. Yes, this is conventional, too, but I'm talking about how we compare the time interval for an event yesterday with the time interval for another event today, even if the events occur far away in space from each other.

Naomi: Good. We agree that we're defining what it is for two noninstantaneous events to last the same amount of time.

John: Yes, and I think these definitions are stipulated, not discovered. Judgments of whether the two intervals are the same will lack truth conditions; there's no matter of fact at stake in virtue of which the judgments can be true or false until we adopt a set of conventions about what we are calling the metric and the standard clock. After we select the conventions, then we can say, relative to those conventions, that the judgment is true that the time between these two ticks is the same as the time between those two ticks.

Naomi: You're being very clear about your position. I've done some reading about this, and I know you've got Poincaré, Reichenbach, Grünbaum, and the majority of philosophers of science on your side, but I'm not convinced. Here's how I look at it. If at some scientific conference all the powerful people who get to decide about time decided to call the Dalai

Lama's heartbeats the ticks of the standard clock, there'd be a big howl, and they'd get complaints that they'd chosen incorrectly.

John: Inconveniently, not incorrectly.

Naomi: Well, it would also be inconvenient, but here's my reason for saying it's more than inconvenience. The Dalai Lama is a spiritual leader with a special inner calm, yet even *his* heart speeds up when he goes jogging. By adopting the standard Dalai Lama heartbeat clock, the present laws that involve time would break down. The period of a pendulum, a weight swinging from a string, would no longer be a function just of the length of its string, but now would also be a function of the Dalai Lama's jogging schedule. If he went for a jog in the park in New Delhi, pendulums in Los Angeles would slow down. How weird is that? It would be scientific chaos because the cause of the slowdown couldn't be found in the processes themselves. That makes the Dalai Lama much too important. There are better choices for the standard periodic process, such as periodic revolutions of the earth around the sun or periodic ticks of atom processes in atomic clocks. The severe inconvenience of choosing someone's heartbeats as a standard periodic process is an indication to me that it's a severely incorrect standard clock. If you choose the correct metric and correct standard clock, you get simple laws; if you don't, then you get scientific chaos.

John: Chaos?

Naomi: All right, I'm exaggerating, but a bad choice can make science much harder to do, and that's a sign the science needs fixing. The truly standard clock we're seeking is the one that makes for the best system of scientific laws. Choosing the Dalai Lama's heartbeats is incorrect because so many of our currently successful scientific laws would break down. Also, we compare choices of standard clocks to find out which choice provides the best explanation of the deviations by the other clocks that might have been chosen. For example, one of our oldest clocks is the daily rotation of the earth; a new noon occurs when the sun is overhead again. But now we have atomic clocks, and the atomic clocks don't stay in synchrony with the earth-sun clock. Which clock should we blame for the deviation? Well, we now know it's the earth-sun clock because the earth's rotation is sped up and also slowed down by the wind hitting mountains. The earth's rotation is slowed in response to the moon's tug on the oceans, and is sped up, but less so, by glaciers melting and letting water go to lower altitude; and all these effects aren't very predictable. If we were to stick with the earth-sun clock as the standard, then the universe's other regular processes—the pendulums, electrical phenomena, radioactive decay—all of these once stable processes would be slowly speeding up unpredictably and for no apparent reason. There'd be all this mystery why the wind and moon and glaciers should be affecting pendulums and radioactive decay. But if instead we choose the atomic clock as the standard, then there's an easy explanation for the earth-sun clock drifting out of synchrony with it. The big point here is that the selection process for the standard clock is really a process of reducing mystery.

John: Maybe atomic clocks are a little better than the spinning earth, but it wasn't incorrect to use the earth clock, just inconvenient. It's not like there is some *fact* of the matter as to whether successive rotations of the earth are really temporally equivalent. You can't snatch one period on a clock and lay it out along another period and see that they're the same interval of time. I think the same goes for distances. We select some stick to be a standard meter and then just declare that it has the same length at all times and at all places no matter where it is transported so long as forces don't bend it or shrink it. It's not like the measure of distance is intrinsic to space itself. It's something we conveniently impose on space to promote science.

Naomi: I disagree. There's a fact of the matter about the standard length and also about the standard clock. We encountered this fact about time when we discovered that the old earth-based standard clock was inaccurate because the tides and the melting ice were causing irregularities in the planet's spin. The choice isn't conventional; it's objective. Choosing an atomic clock instead of someone's heartbeats as the standard clock isn't any more conventional than is choosing to say the earth is round rather than flat.

John: I respect the point you're making, but I still think a declaration of what will be our choice of metric and standard clock has the status of a stipulated definition. Hold on. I just thought of another problem for you. Suppose I want a new clock that I can trust when I'm in L.A. Here's what I do. I buy the clock in Boulder, Colorado, where there's a standard atomic clock that I can use to set my new clock. Then I transport my clock to L.A. How can I be sure that on the trip to L.A. it didn't drift out of synch by speeding up? Suppose I get worried about this so I transport it back to Boulder and find that the two clocks are in synch. Ah, wonderful. Peace of mind. But then I start to worry again. Just because they're in synch when they're together doesn't mean they have to be in synch while they're separated. Maybe my clock sped up on the way to L.A. and then slowed down by the same rate on the trip back, tricking me into thinking the two were in synch the whole time. To gain peace of mind again, I could ship my clock from L.A. back to Boulder for another synchrony check and repeat this every time I start to worry. But don't I at some point just have to say, "Enough with the checking. I'll have faith in my clock"? So isn't that show of faith a source of conventionality, too? I mean for all clocks, not just for my own clock.

Naomi: No, not if there's some well-confirmed theory of physics that implies that your kind of clock transport shouldn't affect clock readings or that implies that transport under those conditions should affect the clock rate by some specific amount that can be corrected for. And there's another reason. There are basic laws of science involving time, such as laws about the period of a pendulum and about a particle's velocity staying constant when no force is pushing on it. If your clock could be used to verify those laws in Boulder but not in L.A., then your clock would definitely be out of synch with the standard clock. It wouldn't be right for you to be stubborn and accuse the basic laws of science of being crude.

John: Have scientists really checked the laws and found that some kinds of clocks stay synchronized and others don't?

Naomi: Yes. So I'd say that if the project of doing science and picking the best overall scientific theories implies that one metric is better than the others and one standard clock is better than the others, then choosing the metric and choosing the standard clock are objective choices.

John: I'm convinced it's an objective matter whether clocks stay synchronized after being transported, but not that choosing the metric and the standard clock is objective. There's my bus stop. Got to get going.

Naomi: OK. Well, I liked our conversation. Enjoy your pay raise.

John: Thanks. Maybe the heat wave will be gone by next week.

Naomi: I'm hoping the ice age returns.

DISCUSSION QUESTIONS

1. What is the difference between psychological time and physical time? Which one is more fundamental, and why is that?
2. Describe Naomi's and John's reasons for why time is or isn't real and objective.
3. Do you suppose that, when bears sense time passing, it feels to them on the inside the way it feels to us? How could you tell, directly or indirectly?
4. Why have some people said Libet's experiments about time refute free will? Why do you agree or disagree?
5. What is correct and what is incorrect about Kant's ideas regarding time?

6. How do you know whether you are an artificial being living in a virtual reality? Take into account your opponent's reasoning on this issue.
7. (a) What makes a clock be a clock? (b) How can you distinguish an accurate clock from an inaccurate one?
8. In selecting a standard clock, why is it better to choose an atomic clock than a solar clock?
9. What features of time are conventional, and what features are not?
10. Discuss this argument. Scientific theories commit us ontologically only to the existence of observables, but time is not observable. There are no circumstances in which time can be observed even indirectly, unlike, say, a very distant airplane in the sky, which is indirectly observable from its vapor trail. Under the right circumstances, we could have observed the plane directly without relying on the vapor trail. A distant airplane is observable; time is not.

FURTHER READING

Blackburn, Simon. *Truth: A Compelling Introduction to Philosophy.* Oxford: Oxford University Press, 1999.
Pages 253–59 contain a brief, clear, and sympathetic presentation of Kant's revolutionary idea about how to understand the relationship between time and mind.

Carnap, Rudolf. *Philosophical Foundations of Physics.* New York: Basic Books, Inc., 1966.
Chapter 8 (pages 78–85) discusses how we compare the length of two intervals that occur at different times, and, more generally, the question of how a good choice of a standard clock will lead to simple laws of physics.

Cowen, Ron. "To Leap or Not to Leap: Scientists Debate a Timely Issue." *Science News* 169, no. 16 (2006): 248–49.
Details of why the Earth clock is slowing compared to atomic clocks. A noon-to-noon day is now 0.002 of a second shorter than it was a century ago.

Damasio, Antonio R. "Remembering When." *Scientific American Special Edition: A Matter of Time* 16, no. 1 (2006): 34–41.
A neuroscientist surveys the brain structures involved in how our mind organizes our experiences into the proper temporal order. Includes a discussion of Benjamin Libet's discovery in the 1970s that some brain events involved in initiating a free choice occur about a third of a second before awareness of the choice.

Dennett, Daniel. "Time and Experience." In *Consciousness Explained*, by Daniel Dennett, 139–70. Boston: Little, Brown and Company, 1991.
Explores how time is represented in the brain and emphasizes his Multiple Drafts model of consciousness.

Le Poidevin, Robin. *Travels in Four Dimensions: The Enigmas of Space and Time.* Oxford: Oxford University Press, 2003.

Chapter 1 (pages 1–12) discusses what it is for a clock to be accurate and whether time's metric is conventional.

Libet, Benjamin. "Unconscious Cerebral Initiative and the Role of Conscious Will in Voluntary Action." *The Behavioral and Brain Sciences* 8 (1985): 529–39.

Many philosophers interpret these experiments as showing that our conscious decisions to act come too late to be the cause of our actions, therefore we have no free will.

Mundle, C. W. K. "Time, Consciousness of." In *The Encyclopedia of Philosophy* 8, edited by Paul Edwards, 134–39. New York: Macmillan Publishing Co., Inc., and The Free Press, 1967.

Surveys philosophical ideas about our consciousness of time.

Nagel, Ernest. *The Structure of Science: Problems in the Logic of Scientific Explanation.* New York: Harcourt, Brace & World, Inc., 1961.

Pages 179–83 discuss what makes a clock be a clock and how we select a standard clock. Nagel argues that the metric of time is not conventional. The discussion emphasizes classical mechanics.

Newton-Smith, W. H. *The Structure of Time.* London: Routledge & Kegan Paul Books Ltd., 1980.

Chapter 7 explores the dispute about whether the choice of the metric and standard clock is conventional. Naomi's criticism of van Fraassen's position that there would be no time were there no beings capable of reason is presented by Newton-Smith on pages 218–21.

Reichenbach, Hans. *The Philosophy of Space and Time.* New York: Dover Publications, 1950.

On pages 116–17, this philosopher of science defends the claim that the choice of metric is conventional because it has the status of a stipulated definition.

Swoyer, Eric. "Relativism." *Stanford Encyclopedia of Philosophy.*

Explores subjectivism, conceptual relativism, and other anti-realist views raised by John in this dialogue. More demanding and broader than the Westacott article below.

van Fraassen, Bas. *An Introduction to the Philosophy of Time and Space.* New York: Random House, 1970. Reprinted in 1985 by Columbia University Press, New York.

Pages 3–7 and 70–80 discuss what makes a clock be a clock, what's involved in choosing a standard clock, and whether the metric of time is conventional. Van Fraassen's argument that there would be no time were there no beings capable of reason is on page 102.

Westacott, Emrys. "Cognitive Relativism." *The Internet Encyclopedia of Philosophy.*

Explores subjectivism, conceptual relativism, and other anti-realist views raised by John in this dialogue. Kuhn, Rorty, and Foucault are leading cognitive relativists. Ernest Gellner has said, "There is no unique truth, no unique objective reality." These issues are pursued at a more demanding level in Swoyer, above.

4

Time Travel and Backward Causation

Naomi: What's doing with your music today?

John: Some tracks from the film *Memoirs of a Geisha*. You know it?

Naomi: Classical Japanese?

John: It's actually by an American, John Williams. He also did all the music for the *Star Wars* films. The cello solos are by Yo-Yo Ma. It was recorded here at UCLA.

Naomi: Oh.

John: Yeah, I wish I had half their talent.

Naomi: I played the piano when I was younger but didn't keep up with it.

John: Me, too. So how was your Physics for Poets class today?

Naomi: We had a little discussion of spacetime.

John: I remember that. It's the big container.

Naomi: A little like that.

John: It's the basic four-dimensional stuff that the theory of relativity says is part three-dimensional space and part one-dimensional time.

Naomi: My students had trouble understanding that time isn't a fourth dimension of space. It's a fourth dimension of spacetime.

John: It took me a while to appreciate that when I first heard it. So I have a question about spacetime that I should have learned the answer to a long time ago.

Naomi: What's that?

John: I'm a little unsure what a reference frame is. It's mentioned all the time.

Naomi: I can help with that. The analyst gets to choose the reference frame for the space. It's the device that assigns locations to all points. Usually this is accomplished by selecting a coordinate system that fits over the whole space and is fixed to some special object at the origin. The Cartesian coordinate system is most commonly used for three-dimensional space. This is the coordinate system invented by Descartes with three perpendicular axes, x, y, and z. The number that helps locate a point along the x-axis is called the x-coordinate, which is a real number indicating how far the place is from the origin. When x is pi, that specifies the point that is pi units away from the origin along the x axis. If the analyst fixes the frame to the airplane you are flying in, then the seats of the airplane are not moving in that frame, though the seats are moving in a frame fixed to the earth.

John: What about spacetime?

Naomi: Spacetime is a four-dimensional space, and commonly its coordinate system is like Descartes' except that a fourth axis is added perpendicular to the other three; this is the time axis, with t for the time coordinate. If the space has especially unusual properties, then the analyst may choose a different coordinate system. For some spaces no single coordinate system will fit over the whole space.

John: All this is a lot simpler than I thought. Thanks.

Naomi: Happy to have a little success. I didn't have as much success in my own class. Sort of frustrating. I'm reminding myself that this is a day I'll never get back. Hey, that reminds me. Today was your time travel day, wasn't it?

John: Yeah. Can you see me traveling along the time dimension? I'm moving right now at a rate of one second per second.

Naomi: You call that traveling?

John: Yes, to the present. What would you call it?

Naomi: Existing. How about real time travel that escapes the present?

John: I can drive east from L.A., and when I cross into Arizona I'll enter the Mountain time zone and jump into the future by one hour.

Naomi: That's not real time travel either.

John: It depends on what you mean by "real."

Naomi: Suppose right now I'm thinking back to the time I got lost walking home after the first day of sixth grade at my new school. Have I just now traveled back in time to when I was in the sixth grade?

John: Sure, your mind travels back to an earlier time.

Naomi: You're taking all this too literally.

John: It's how people talk.

Naomi: True, but it's all metaphorical.

John: There's another kind of time travel.

Naomi: What's that?

John: Using a refrigerator.

Naomi: Right, and you're having a sale on refrigerators today, but today only. Will you sweeten the deal by including a free two-quart spaghetti strainer and clam steamer?

John: I'm serious! Here's how it works. I'm quick-frozen, then thawed out a century later. I awake in the future, as young as I am now. OK, OK, roll your eyes if you want, but it's future time travel if you think about it hard enough. I'd call it "biological time travel."

Naomi: That's not time travel either, at least the way I think of time travel. Do you have some sort of definition of the term?

John: Here's the definition Mehlberg recommended in class. You travel in time if the difference between your departure and arrival times judged in the surrounding world doesn't equal the duration of the journey you've undergone personally—that is, as judged by your own clock.

Naomi: Let's see. Your personal time is what the theory of relativity calls your "proper time." It's the physical time shown on a clock that travels with you. So if your personal clock gets out of synch with the external clock in a reference frame fixed in the surrounding world, then you've traveled in external time, assuming both clocks were working properly. OK, I understand the definition. Let's see how it works. Crossing time zones won't be time travel. Any switch of zone is a switch in the external reference frame. The definition assumes you first pick an external reference frame and stick with it. Your time travel by zone crossing is really just an equivocation on the phrase "reference frame fixed in the surrounding world."

John: OK, but what about time travel by remembering the sixth grade, or daydreaming that I'm a knight in a medieval castle?

Naomi: Your personal clock doesn't start showing medieval time just because you start dreaming you're a medieval knight.

John: I see. OK, forget that. But how about time travel by biological freezing?

Naomi: Your personal time can be measured on a clock that doesn't freeze.

John: Can't a real personal clock freeze when you freeze?

Naomi: Any real clock can freeze, but reference frames are imaginary laboratories, and their clocks don't freeze. If two hikers are ten feet from each other along a path during a rainstorm, the ten feet itself doesn't get wet. What gets wet are the hikers and the path they're on, not the distance itself. Does that make sense?

John: I see. So I guess I should say my freezing would slow down my biological aging but not slow down time.

Naomi: Yeah.

John: I didn't realize Mehlberg's definition of time travel ruled out so much. Today's class topic was about time travel, and we used that definition. We didn't talk about all those things you and I were talking about. Too bad. They were on my mind during the whole discussion. Instead, we talked a long time about these two twins who synchronize their clocks. One climbs into a spaceship with his clock, zooms off at very high speed, returns, lands back on Earth, and compares his clock with his twin sister's clock. She's aged more, and her clock shows more elapsed time than her brother's. That is, the brother's personal time is slowed compared to the external time measured by his sister. The closer the spaceship gets to the speed of light, the cosmic speed limit, the more out of synch the two clocks get, yet both clocks are giving correct readings for their own personal times. Mehlberg said this is because there's no "true" time that is reference frame free.

Naomi: More people would visit the future—maybe to become younger than their children—if their spaceships weren't so slow and the travel wasn't so dangerous. Science fiction films leave out the little detail that no walls of a spaceship would be strong enough to protect them if a speck of space dust ripped through at almost the speed of light.

John: But if it weren't for that safety problem, the parents could time travel to a future where they are the same age as their child.

Naomi: Yes. Einstein's main original idea about time is that there may be "good cop, bad cop," but not "good clock, bad clock." A clock isn't necessarily bad simply because it doesn't read the same as a clock it's been synchronized with. Two perfectly running clocks can differ if they've been moving relative to each other. That's why how long an event lasts depends on which clock is being used, unlike in a world obeying Newton's laws where moving clocks always have to tick just as fast as the stationary ones. There's no time travel in Newton's world.

John: So if the world obeyed Newton's laws, the two twins would stay the same age?

Naomi: Yes, but in our real world with Einstein's laws, the space traveler goes into his twin sister's future, judged from her clock, but he's always in his own present as judged from a clock fixed to himself.

John: What about time jumps, like in science fiction stories?

Naomi: In those stories, the time traveler leaves the present and pops into the future—surprise, surprise! That's not allowed by the theory of relativity. All travel has to be continuous in any frame. Suppose the travel begins in the year 3000 and lasts five years by earth clocks, but the speeding traveler comes back in 3005 having aged six months on his own clock. Earth astronomers can watch him continuously for all five years. He won't "jump" into their year 3005. He will slowly tear six sheets off his monthly calendar during those five years that he's racing around space at close to the speed of light. When his six months of time stretches out to five years like this, it's called "time dilation."

John: I wonder how we ever keep several clocks synchronized with each other.

Naomi: We don't. When we say, "OK, everybody, synchronize your watches," we can't rely on the watches staying synchronized. They'll go out of synchrony if one of them starts moving relative to the others. Time dilation destroys synchrony.

John: OK, but what I still don't understand is this situation with the twins. Suppose the twin in the spaceship comes back to Earth in 3005 Earth time and notices it's only July 3000 on the clock in his spaceship. Why can't the sister who stayed on Earth just step into her brother's spaceship and instantly be transported from 3005 back to July 3000? Isn't that a time jump?

Naomi: Yes and no. She won't be back in the year 3000 in her own frame, only in her brother's. You can get time jumps only by abruptly equivocating and shifting the frame you are using to make time judgments—like when you suddenly decide to shift from using Pacific Standard Time to Mountain Standard Time. Time dilation is real time travel. But it works only one way, to the future, not to the past.

John: Oh, I see now, but you've just spoiled my next issue of *Science Fiction Magazine*.

Naomi: Sorry for that, but reality is pretty bizarre, too. In *general* relativity, there's a second kind of time dilation; it's due to high gravity. Did you discuss that second kind?

John: Yeah, Mehlberg said that because of gravity being stronger in a first-floor apartment than in the penthouse I'd live longer on the first floor. The time dilation isn't very significant, but it would be if I lived near a black hole. If I fell directly into a black hole, the high gravity would suck me into its center quickly as judged by my own clock, but as judged by a clock back on Earth it would take an infinite amount of time to fall in. Imagine that! As far as the Earth people are concerned, I'd have traveled to the end of time. To infinity and beyond!

Naomi: [laughs]

John: What would you ask some guy who said he's a time traveler from our future?

Naomi: I'd ask him whether he's taken his pills today.

John: Good question, but I'd say it's impossible to travel to our time from the future. For one thing, if it were possible we should have seen time travelers by now, but who's seen one?

Naomi: Maybe it's possible, but our times are so boring that nobody wants to visit.

John: All right, but if I'm wrong, then I'd like to travel from here back to improve history. You know, stop wars before they start, prevent Abraham Lincoln's assassination, that kind of thing.

Naomi: Did you talk about the paradox involved in trying to do that?

John: Yes, I go back in time and prevent my grandfather from having children. Then I'm not born. It's the Grandfather Paradox.

Naomi: So, if you do go back, you'd better not interfere in what has already happened. You can't save Lincoln from assassination, at least not in this universe, because he has already been assassinated. If time travel to the past is possible, it's got to be that time travelers can go back and affect history only in the sense of *participating* in history, but not changing it. A similar point holds for travel to the future. You can enter a spaceship and speed up so that you travel to the future where cancer is cured, but you can't travel back with the cure to today or any earlier, even if backward time travel is possible. You can't make a world that has no cure for cancer at a certain time also have a cancer cure at that time.

John: OK, but how about my bringing back the cure to some later time such as tomorrow?

Naomi: That's possible. . . . There's a way you can travel back in time and kill your grandfather and save Abraham Lincoln. Travel to a parallel universe and save them there.

John: Parallel universes seem to be a lot like the possible worlds we talk about in metaphysics class, and the problem there is whether the other world contains Lincoln or just a counterpart of Lincoln. I want to save the actual Lincoln, not just somebody who looks the most like Lincoln in that world.

Naomi: Don't hold your breath.

John: I know. It's just wishful thinking. The student in my seminar described a more realistic scenario about travel to the past. It's about going back in time and becoming Josef Stalin, the dictator. He was twenty-one years old in 1900. Suppose time machines are invented in 2080. In that year, some guy, let's call him "Sam," decides to assume the identity of Stalin. He knows his Russian history, speaks fluent Russian with the right accent, is twenty-one years old, and looks and acts like Stalin did at twenty-one. Sam enters the newly invented time machine, goes back to 1900 (don't ask me how), secretly murders the young Stalin, then starts calling himself "Stalin." Sam's secret remains safe, and he eventually becomes the dictator and does all those horrible things we've read about. He doesn't do anything different than what we know from the history books; there's no changing the past, just participating in it. The question for metaphysics is this: is it remotely possible that this is a true story about Russian history?

Naomi: I think it's possible as far as logic and the theory of relativity are concerned, provided Sam doesn't suddenly pop into 1900 but gets there by going forward in his proper time until he ends up back in 1900 as shown on the young Stalin's clock.

John: But there's the issue of what makes a person be the same person from one time to another. Because we already know Stalin died in 1953, Sam will have to die in 1953, many years before he's born. Some metaphysicians would say this is metaphysically impossible.

Naomi: What do you mean by "metaphysically impossible"?

John: Logically inconsistent with metaphysical truths, such as the truth that persons are born before they die.

Naomi: I can't see how you'd establish this metaphysical truth except by declaring it's true by definition or by how we use the word.

John: OK, that's what I declare, or assume, or whatever.

Naomi: Since the Stalin story doesn't violate physics, maybe it shows that these metaphysicians ought to change their definition of "person."

John: I'd rather not do that if I can help it. Maybe if I tell you some of the other odd consequences of time travel, you'll reject time travel instead of metaphysical truths. If we accept the Stalin story, then we might have some very strange time travelers visiting us. I remember, when I was a child, my mother and I saw a burglar peering into our kitchen window one night. The burglar could really have been me visiting from the future!

Naomi: Uh . . . maybe.

John: And it gets worse. Suppose I do go back in time to visit my child-hood self in the kitchen that night. Some metaphysicians say this would be an example of backward causation because my getting into the time machine is causing the kitchen event at an earlier time.

Naomi: Wait, backward causation is different from backward time travel, right?

John: Yes, time travel requires something to be transported from one time back to another time; backward causation doesn't. Or at least it doesn't if you allow there to be a causal influence without there being some object transporting the cause to the effect.

Naomi: Surely you'd want to allow this kind of causal influence; it's called "action at a distance." When Isaac Newton came up with his theory of gravitation, he used it to show that earth's gravity causes the moon to orbit the earth. This influence from earth to moon was instantaneous, an action at a distance. It wouldn't be fair to rule out Newton's theory simply because he violated a philosopher's definition of "cause" that didn't allow action at a distance.

John: I think nearly all metaphysicians would agree to allow action at a distance as a possibility; it would be up to specific scientific theories to rule it out. But isn't any time travel to the past a case of backward causation because your entering the time machine would be causing earlier events such as your exiting the time machine?

Naomi: Yeah, it's always backward causation.

John: But backward causation can be ruled out for all sorts of reasons, so time travel to the past is impossible. First off, backward causation of some event now by an event in the future is impossible because the future doesn't exist. So future events can't be causing anything.

Naomi: Future events don't exist *now*, but you can't say that future events don't exist at all.

John: This issue about whether the future exists is on the syllabus a few weeks ahead, so I'll hold off my response until then.

Naomi: OK. Do you have any other problems with backward causation?

John: Yeah, because it's true by definition that causes are earlier than their effects.

Naomi: Are you saying that time travel implies backward causation, but backward causation is impossible because it violates the definition of "causation," so time travel is impossible?

John: Right.

Naomi: If I were to prove to you that Einstein's theory of relativity permits time travel, would you say this gives you a proof his theory is false?

John: No, I'd just say that even if his theory permits time travel, time travel never actually occurs. His theory permits too much.

Naomi: OK, I'm happy you aren't trying to refute Einstein. Well, it definitely *is* weird to have the effect first and the cause second, but to me it's not as ridiculous as there being a bachelor who's married. Maybe backward causation is only violating common sense and not violating logic or any definition we need to preserve. On the other hand, I don't think it's ridiculous to define cause the way you do, especially if we are sure we have enough other knowledge about causes that indicates this would be the right definition. But we don't have this knowledge, and I don't want to start with any definition that wins something by begging the question.

John: Don't we know, from the very meaning of the word, that time is a continuous line? I mean a noncircular line.

Naomi: Like a line, not is a line.

John: OK, like a line.

Naomi: You can't be so sure that the word "time" has this meaning. You can't trust ordinary dictionaries when it comes to something like this, even if you could find this in a dictionary, which I doubt. Besides, the ancient Greek Stoics believed time was circular, and they knew what the word "time" meant, didn't they?

John: Maybe not. Look, once you allow time travel you've got circular time, and I don't know what "happens earlier" means when time goes in a circle. In circular time, any pair of events would have the strange property that each of them happens before the other and also after the other, which is way too weird for me to think about. All novelty would be sucked out of the universe because every event would keep happening over and over. It's a terrifying prospect. You can give a mathematical description of something or other that isn't linear, and you can call it "time," but it won't really be time. There have to be some features of time we hold onto in our definition of the word "time." This is a battle of liberals versus conservatives. We already talked about being liberal enough to change the meaning of the word "person." Now you're talking about changing the meaning of the word "time" to allow it to go in a circle. You're giving birth to an oxymoron.

Naomi: Who should be doing the defining—experts or ordinary people who talk the way they want and don't think about the deep issues?

John: It's not that clear cut.

Naomi: OK, I agree with you on that, but concepts can grow as we learn how to use them to solve problems and deepen our understanding. If they grow too far away from the old meaning, then maybe we do need to pick a new word, or else say that now we're using the term in its technical sense. We need to keep an open mind about the possibility of allowing time to be like a line or a circle or the letter Y or whatever. The same goes for backward causation. I guess I look at things this way. You could never have learned that water is H_2O just by examining the concept of water, and you can't learn what causation is by examining the concept of causation. You need to let our discoveries drive changes in the meanings of our concepts.

John: Sometimes we want to make our concepts precise, so we change them a little, like when a judge makes the concept of "theft" a little more precise by deciding whether to call me a thief for voluntarily shoveling the snow off my neighbor's driveway while keeping that snow for myself. But when you say time is circular or causation goes backward, you aren't just making the concepts precise; you're changing the subject.

Naomi: I don't believe the concepts are that sharply defined for you to say the subject is changed. Besides, there's no precise way to distinguish between what we contingently believe about time and what is true necessarily. I think of concepts more like muscles; the more you use them, the more they develop and become better able to deal with problem cases. I'm tempted to say that the idea of causes usually preceding their effects is some analytic kernel of our concept of causation, and the idea of time being linear, at least for short times, is some kind of kernel of our concept of time order that we might hold onto, but even then I'm unsure. Anyway, can we agree to disagree on this point, and then consider what else is involved with circular time?

John: I guess so. OK, suppose our personal time can go in a circle. Here's the problem. In a one-day loop, your having breakfast causes you to have lunch, which causes you to have dinner, which causes you to have that same day's breakfast. That's absurd, so there can't be causal loops.

Naomi: Strange doesn't mean impossible.

John: Absurd does. Besides, there are even more reasons why backward causation and circular time travel are impossible. It's got to do with bilking. Bilking is like refuting, but it's a kind of action. Like when you act to refute claims of backward causation by observing the supposed earlier effect and then preventing its later cause from occurring. It seems intuitively obvious to me that this kind of bilking will succeed and then the whole universe blows up.

Naomi: Blows up?

John: Not literally. Here's what I mean. Take a case where Sam the time traveler arrives back in Russia in 1900. Let's temporarily accept your idea of backward causation and say his arrival in 1900 is caused by his later entry into a time machine in 2080. If there is this time travel to the past, then the future isn't so open anymore. It's too oddly constrained because the universe's events in 2080 must, absolutely *must*, allow Sam to enter the time machine. If he can't enter, then the past is inconsistent, which is impossible. Everything Sam does as a time traveler is already built into the past and must happen when viewed from 2080. Nothing the bilkers do can prevent Sam from getting into the machine. If they try to poison him, the poison must fail to work. If Sam were shot by a sniper from the People's League to Combat Time Travelers, the bullet couldn't possibly kill him. Looking back from the year 2081, it will appear as if the universe conspired to protect him and ensure that a predestined event occurs—namely climbing into the time machine. So, in 2081, we'd see that there's no free will and the future is constrained to never allow bilking. That's absurd. The bilking has to succeed. If we have free will, we can stop Sam getting into the time machine. But we do have free will, so we should conclude backward causation can't happen.

Naomi: That's a clever argument, but I think it's an overreaction. If the bilking fails, this would simply be surprising, but we could hold onto free will and just say that we never were free to do the impossible. Killing Sam was impossible in 2079 because he had already safely made it back from 2080 to 1900. There's no free will problem because it's unfair to demand that we have the freedom to do the impossible.

John: I guess that sort of saves free will—barely. But I'd prefer to say it's what happens in 1900 that is accounting for all those later attempts on Sam's life to be botched. Now look at what this does for your beloved backward causation. It's really the earlier events that are influencing the later ones, and there's no backward causation after all.

Naomi: Well, let's see, I don't think you've got a general refutation of backward causation here, only an insight that unusual coincidences occur when there's backward causation, like failure of bilking. It seems to me that backward causation is possible, but because it occurs so infrequently, we tend to rule it out when we're searching for causes of phenomena, and we try desperately to explain things with forward causation. But actually the backward causation is just unlikely, not impossible.

John: I assume you think causation is some objective feature of nature that isn't just something our mind merely projects onto nature, don't you?

Naomi: Yes.

John: If you think this way, I'd like to see you define causation between two events without assuming you already know which of the two came first. I wouldn't be surprised if your definition rules out backward causation.

Naomi: OK, that's fair. I'll have to think how I can be clear about what it means for causes to make their effects happen. Hmm . . . It isn't obvious to me how to do it, but I definitely don't think causation is *subjective*.

John: I have another good reason why time machines are impossible.

Naomi: You do? By the way, in class, did you talk about what a time machine would be like that could take somebody to their own past?

John: The question never came up.

Naomi: From my reading, nobody is about to get a patent soon on one. But anyway, fire away with your new reason why the machines are impossible.

John: The machine would allow us to get information for free, and that's not possible. For example, let's say you want to know how lasers work and how to build one. So tomorrow you read about this in a technology encyclopedia and take notes. You enter a time machine with your notes, go back, and give them to someone who then builds the world's very first laser by following the instructions in your notes. After this first laser is built successfully, someone describes it by writing an article in a technology encyclopedia. It's the same encyclopedia you will read tomorrow. So here's the problem. Where did the idea of how to build a laser come from? You didn't think of it; you got it from the book. The inventor didn't think of it. He got it from you. See the problem?

Naomi: Yes, but the laser was invented. . . .

John: I know the laser wasn't really constructed this way, but my story is consistent and could describe the discovery of future pieces of technology that actually will be constructed by following notes.

Naomi: What a snarl. I'd say this is your best argument against time travel and backward causation. I can think of only one way to wiggle out of this paradox—declare it's possible to have some free information even though most information has a creator.

John: I'd prefer adopting an axiom or first principle that says free information is impossible, and that's why time travel is, too.

Naomi: That's drastic.

John: What's drastic is biting the bullet and allowing free information. I think you're on the defensive here, but I've got another reason to oppose time travel.

Naomi: Where's my tranquilizer gun? You're really excited about this issue, aren't you?

John: Yeah, I am.

Naomi: What's the other reason?

John: Suppose we're in an era when time machines have been invented. One day at noon I give you the instructions to look for a certain coded signal that I'll be sending you from my time machine later that day at two o'clock. You have agreed to follow the instructions whatever they are. In that signal, you're instructed to blow up my time machine right away if my signal ever arrives, but otherwise you're instructed not to harm the time machine. Now, the question is this. At two o'clock can I use the time machine to send the coded signal back in time to you at one o'clock?

Naomi: Well, let's see. If you can send the signal, then the instructions will lead me to destroy the time machine before you step into it, but then you can't send the signal. And if you can't send the signal, then the machine won't blow up, so you can send it. I guess the conclusion is that you can if and only if you can't. Wow!

John: Not quite.

Naomi: Why not?

John: Our class came up with three ways out of this paradox. We could say (1) the universe conspires somehow to keep people from obeying these sorts of destructive instructions or prevents them from detecting the signal once it's been sent, or (2) the signal that is sent into the past at two o'clock must go so far back in time that it never survives to make it back to one o'clock, or (3) signaling the past is always impossible.

Naomi: I like 1 and 2.

John: I like 3.

Naomi: OK, let's flip for it. Heads, I win. Tails, you lose. Your choice.

John: [laughs]

Naomi: Let's make whatever choice causes the least harm to the rest of our body of knowledge. You know, minimize our losses.

John: I wonder if we could get agreement on which changes cause the least harm since I'd be more harmed by some changes than you would. Well, got to go. I'm happy the bus is getting near my refrigerator. I missed lunch.

Naomi: OK, I hope you don't see your older self peeking at you through the kitchen window.

John: If I do, I'll let you know a half hour ago.

Naomi: Right. See ya.

DISCUSSION QUESTIONS

1. Explain why switching to Daylight Saving Time isn't a kind of time travel.
2. In principle could you travel to the future and read about what caused your death? Why?
3. Summarize all the reasons offered by John against time travel, then rank them from strongest to weakest and defend your ranking.
4. (a) How does John define "time travel"? (b) Does his definition cover everything that deserves to be called "time travel"? (c) What situations does it rule out that some other people might call "time travel"?
5. Suppose one day your friend Kyle says to you that he's actually a time traveler from Jerusalem who left there in AD 55 and arrived

here in a few years. To prove it, he shows you a piece of wood that he says was part of the cross on which Jesus was crucified just outside of Jerusalem. You think, "I might be able to check on Kyle by doing a radiocarbon dating test of the wood." (a) If Kyle were really telling the truth, what date should you get from the test—that the wood is, say, 20 years old or 2,000 years old? (b) Do you think the results of your test could provide helpful evidence about whether Kyle is lying or delusional? (c) Devise another way to test Kyle's claim.

6. Assuming time travel is physically possible, can you go back in time of your own free will or are you forced to go because you already did?

7. Does stepping into a time machine cause earlier events?

8. When we speak of personal identity we are talking about what constitutes the person from one time to the next, not how we *recognize* the same person again. (a) What constitutes being the same person from one time to another? (b) Could a friend of yours die before his or her birth? (c) Could you be the same person you are now but have had different parents? (d) When did you start being a person?

9. Consider this supposed case of backward causation. Distant hunters act bravely during the hunt for food because after the hunt the tribal chief performs the hunting dance. How would you go about bilking this claim of backward causation? What should you conclude if you discover that all your bilking attempts fail for some reason or other?

10. What is the solution to the paradox John presents about sending the coded signal instructing Naomi to blow up his time machine right away?

FURTHER READING

Davies, Paul. *How to Build a Time Machine.* New York: Penguin Books, Penguin Putnam Inc., 2001.
   A popular exposition by a physicist of the details behind the possibilities of time travel.

Deutsch, David, and Michael Lockwood. "The Quantum Physics of Time Travel." *Scientific American* (March 1994): 68–74.
   A physicist and philosopher investigate the puzzle of getting information for free by traveling in time along a world line that forms a closed loop, and they argue that there is no good reason to disbelieve in the possibility of travel to the past.

Dummett, Michael A. E. "Bringing about the Past." *Philosophical Review* 73 (1964): 338–59.

An early discussion of bilking involving the tribal chief who performs a ritual dance in order to produce bravery in his hunters.

Hawking, Stephen. *A Brief History of Time: The Updated and Expanded Tenth Anniversary Edition.* New York: Bantam Books, 1998.

In chapter 10, Hawking discusses why he believes time travel to the past is impossible. Chapter 2 explains what spacetime is and why we should believe in it.

Horwich, Paul. *Asymmetries in Time: Problems in the Philosophy of Science.* Cambridge, MA: The MIT Press, 1987.

Chapters 6 and 7 of this philosophical monograph are devoted to backward causation and time travel. Bilking is discussed on pages 93–99.

MacBeath, Murray. "Who Was Dr. Who's Father?" *Synthese* 51, no. 3 (1982): 397–430.

Explores the strangest consequences of allowing time travel.

Mellor, D. H. *Real Time II.* London: Routledge, 1998.

Argues that backward causation is incoherent. Also, defends the tenseless theory of time that will be introduced in chapter 6 of this dialogue. Difficult reading.

Thorne, Kip S. *Black Holes and Time Warps: Einstein's Outrageous Legacy.* New York: W. W. Norton & Co., 1994.

Chapter 14 describes how to use a wormhole to create a time machine.

5

Time's Origin, and Relationism vs. Substantivalism

John: Whoa! What happened to you?

Naomi: I went horseback riding this morning and rode for a whole five seconds, then landed in a bush and scratched my face. When I was flying through the air, time stopped.

John: I'll bet. Anything broken?

Naomi: Just the bush. My shirt looked like it was eaten by gerbils, but I'm fine now except for some bruises and scratches.

John: Good. Hopefully you'll heal fast.

Naomi: Thanks.

John: How'd your day go? Any good discussions in the physics class?

Naomi: We talked about magnetism, and whether magnets can affect our body.

John: Do they?

Naomi: Not very much. Technicians in physics labs are around very high magnetic fields all day, and they don't have any problems. A doctor's MRI scan uses very high fields, too, with no problems. Electromagnetism is a different story. Intense light can blind us and burn our skin. X-rays are worse. The highest frequency of all, the gamma rays, blast us like Swiss cheese. But electromagnetism isn't the same as magnetism.

John: Did you have fun talking about all this?

Naomi: Yes, teaching is a great way to make a living. I can imagine doing it for eternity.

John: The trouble with eternity is that there's no telling when it will end.

Naomi: Something like that. So how was your time seminar today?

John: Interesting. We looked at time all the way back to the dawn of time.

Naomi: What makes you think there was a dawn of time?

John: I'm not sure there was. Some students said the astronomers believe the dawn of time is when the first event occurred 13.7 billion years ago at the beginning of the big bang. But now that I think about it, 13.7 in what reference frame?

Naomi: Good question. In a special reference frame that physicists use for all times involving cosmic events. In this frame the faint microwave background radiation left over from the big bang is about the same in all directions. Measuring time of cosmic events in this frame isn't significantly different from measuring them in a frame fixed to the earth because the relative speed of the two frames is only one-thousandth the speed of light. That's why the special frame usually isn't mentioned.

John: I see. Well, the universe had to exist eventually; it's just a question of when.

Naomi: I don't think there had to be a universe at all.

John: Of course there did.

Naomi: Why?

John: There can't be just nothing.

Naomi: I don't see why not. It would be exciting if some day we figured out that our universe has a certain characteristic or law that implies that the universe necessarily had to exist, but I'm betting against this. And even if we found the law, we'd immediately want to know why we have a universe with the law instead of without it.

John: Are you suggesting our universe just poofed into existence one day?

Naomi: Yes. How can we rule this out?

John: If a box of pistachio nuts showed up on my front porch with no note on the box, would I say, "Oh, maybe this just poofed into existence"? That's no method to adopt when you're trying to find explanations. And now you're suggesting this poof theory as the explanation of the origin of everything?

Naomi: I'm not very excited about this myself, but maybe we all have to live with this brute fact. There just is a universe that started with a big bang, and there's no reason for it—no scientific explanation, and no non-scientific explanation.

John: I'd prefer to say there's a perfectly good explanation of why there's something rather than nothing, but humans aren't smart enough to understand it.

Naomi: OK, maybe that's the human condition. I don't know. I have the same desire you do to find some sort of explanation. Hey, here's a way for the universe to poof into existence from no earlier universe. If there's backward causation, then maybe the universe's coming into existence with a big bang was caused by some state in its future.

John: You know what I think about backward causation.

Naomi: Yes, but I still think it's a physical possibility that can't be ruled out. Or how about this? Maybe the universe created itself.

John: That can't be right. We've never seen anything else create itself; it's metaphysically absurd.

Naomi: I don't think it's absurd, just surprising. After all, there are serious scientific theories on how the birth was just random. But the theories are esoteric and haven't been tested and confirmed.

John: OK, you have your reasons why the universe might have popped into existence from nothing, but I have my reasons for why it had to happen very differently. The universe had to exist in order for it to produce you and me who are talking about why it exists.

Naomi: That doesn't smell right.

John: Having olfactory prowess isn't an argument.

Naomi: I meant that's no explanation.

John: Sure it is.

Naomi: That's a flimsy excuse for an explanation.

John: But it might be the right explanation. OK, let's try another explanation. Anything that's logically possible is actual in some possible world or other. There's a possible world just like ours with rocks and air and people, but without Abraham Lincoln ever having been a U.S. president. There's another possible world in which some of our scientific laws are false. So the reason our specific world exists is that all possible worlds exist.

Naomi: That would explain our world provided we knew that all possible worlds had to exist, but if the only reason to think all possible worlds exist is that it gives an explanation of why our world does, then this is circular reasoning.

John: Ah, but there are independent reasons to believe in possible worlds. You probably believe it's true that Lincoln might have survived the assassination attempt. What makes it true that he might have? It's that he actually did survive the assassination attempt, but in another possible world. See? We need possible worlds to make sense of terms like "might have survived" and "could."

Naomi: Oh, OK, I see why you might think possible worlds are helpful, but you should give those worlds as-if standing and not take them literally. For one thing, to decide whether something is possible, you can't search through all the possible worlds to see. Look, there are two questions we want answered. Why is there something rather than nothing at all, and why this something rather than some other? Your assumption that all possible worlds exist will answer the first one, but I don't find it to be a very satisfying answer because it doesn't answer the second one. Explaining Lincoln's assassination by saying he had to be assassinated in some world or other doesn't help me understand his assassination. A real explanation should tell us about conspirators and who pulled the trigger and the motive. That's why I'd prefer saying nobody knows the answer.

John: OK, I'll focus on our world. God made our world, probably by using a big bang.

Naomi: I don't think adding God into the picture explains anything either. We've got this mystery of the origin of our universe, and this mystery doesn't get solved by adding another mystery with the remark that God made it. Maybe he did; maybe he didn't, but if he did, then tell me why he did and how he did, so I can see that your explanation is better than just saying nobody knows.

John: God works in mysterious ways, they say.

Naomi: I can't disagree with that.

John: Maybe he did it with miracles. How's that?

Naomi: Not good enough. I think that's just a fancy way of saying nobody can know how it happened.

John: It's much more intuitive to accept that God exists necessarily and that somehow his Nature requires him to create our universe than it is to accept that the universe might have poofed into existence accidentally. This necessity is a step toward understanding why there's something rather than nothing. I at least have a first step; you don't.

Naomi: It's better to stand still than to take a first step over the cliff.

John: Terrible analogy, but I think we've made progress just by getting clear about our disagreement. I want to make another point though. If time doesn't go in a circle, then the universe must either have a first event or have existed infinitely back into the past. That's just a matter of logic, and if so . . .

Naomi: Hold on. Don't limit us to those two choices.

John: What other choice could there be?

Naomi: There's a way the universe might have a finite past and also no first event.

John: Now what are you getting at?

Naomi: Think about the time line being like the natural ordering of the positive real numbers with each instant associated with its own positive real number. Then maybe looking back to the big bang is like following the positive real numbers back to ever smaller positive numbers without ever reaching a smallest positive one. By analogy, for each event there would be an earlier event, yet there'd be no starting event. If that's possible, then it's a mistake to say the universe must either have a first event or have existed infinitely back into the past.

John: OK, I'll stop saying it's a matter of logic that there was a first event if there was a finite past. My real concern is how we can know it's finite. What stops there being a pre–big bang?

Naomi: Many physicists say the big bang had a first event and was the beginning of time, but they aren't very sure because this classical big bang theory is based on the assumption that the current universal swelling of space can be projected all the way back to zero volume, using the fact that gravity is the only significant force affecting the expansion. Yet physicists universally agree that the projection must be unreliable for all times less than 10^{-43} seconds after the big bang began. Before this, the volume is so small and the density is so great that quantum effects are as significant as any gravitational effects. Because there's no successful theory yet that incorporates both gravity and quantum theory, scientists aren't sure what to say about time any earlier than this.

John: Where science leaves off, metaphysics takes over.

Naomi: Science hasn't left off. Physics has many theories about what happened at and before the big bang. The problem is that the theories aren't tested.

John: Kant gave a metaphysical argument that you can get a contradiction regardless of whether you assume the universe came into existence a finite time ago or an infinite time ago.

Naomi: What sort of contradiction?

John: Well, he begins by assuming along with Isaac Newton that time exists independently of events in time, which is an assumption I'd agree with but you probably wouldn't. In his first contradiction, he supposes the universe has always existed. But if so much time was available, then why did it take so long to reach the state it's in today? There's no good reason. Also, a completed infinity of successive states of the universe would have occurred by now. But that contradicts our knowledge that an infinite process can never be completed. So we've arrived at a contradiction. It's like if someone were to say they've been counting the negative integers and today they finally reached negative one. Isn't that absurd?

Naomi: Maybe, but where'd we get "our knowledge that an infinite process can never be completed"?

John: We know it from the very definition of what "infinite process" means.

Naomi: I don't like this old-fashioned definition. If we use a contemporary definition of infinity, one that allows completed infinities and not just potential infinities, then I don't think Kant can deduce his first contradiction.

John: That's what Mehlberg said. He asked us to hold off on this controversy until we discuss how Aristotle solved Zeno's paradoxes because the controversy is central there.

Naomi: I'll be patient.

John: Kant's second contradiction arises from the assumption that the universe came into existence a *finite* time ago. But "Why then?" he asks. Since time existed back before this origin event, there'd be no good reason why the universe came into existence then instead of at some other time, which is absurd because there are good reasons for everything.

Naomi: Clever, but Kant is jumping to conclusions. How is he justified in saying time existed before this origin event?

John: You've brought up the main criticism, but I think his second contradiction contains a germ of truth. Time has its properties noncontingently. It's a matter of metaphysical necessity or logic that for any time there is an earlier time. Here's why. Let the period P be from now back to the big bang event. Either there were dollars somewhere prior to P or there weren't. In either case there must have been a period prior to P during which there were or weren't dollars. See what I mean?

Naomi: What I see is that we use language this way, and you're trying to encode this casual language use into some principle of logic or metaphysics in order to win an argument about the time existing before the first event in time. The logic of possibility doesn't require a bank to be bigger in order to handle the possibility that there might have been more money in the vault than there is. You're trying to lure me into assuming, as a matter of the logic of time-talk, that time has to be as "big" as any possibly long process back in time. But why accept that? Time needs to be only as big as it actually is, and saying this doesn't rule out the possibility that it might have been bigger. It's logically possible that there are times before 13.7 billion years ago. But you can't say the times must be actual in order for them to be possible.

John: If something could have happened earlier, then there has, absolutely has, to be a time at which it might have occurred earlier.

Naomi: Here's my guess why you keep thinking this way. Maybe you're overly influenced by the idea that we use numbers for times. We assign the number zero to some event, the birth of Jesus or the entrance of Mohammed into the village of Medina, and then with this zero time we start talking about times before that, and so on, assigning negative numbers to all these times. Well, there are infinitely many negative numbers available, and you've jumped from this to saying there are that many past times. This jump is confusing time with the representation of time.

John: I don't know if that's what I'm doing. What I'm more sure I'm doing is trying to convince you to accept Newton's substantivalist theory of time.

Naomi: I've heard of it. Time is a substance like trees and copper?

John: Yes, but those are material substances. The substantivalist theory says time is an immaterial substance that exists even when there are no changes, or, worse yet, even when there are no objects at all. The relational theory—the one you're attracted to—implies there can't be time without change. There's a similar dispute about space. The substantivalist theory of space treats space as an immaterial container for objects, but it exists with or without them. The relational theory implies there's no container without contents. Think of what you get when you take away all the objects from the universe—all the matter and energy. The substantivalist answers, "Empty space." The relationist disagrees and answers, "No space; not even empty space." In parallel with those answers about space, what do you get when you leave the objects alone but take away all their changes? The substantivalist says, "Empty time with nothing happening," but the relationist says, "No time at all." The substantivalist theory is closer to common sense. The relational theory is a negative theory because it implies that time as most people understand it doesn't exist, but it's the most popular theory among philosophers and scientists, and I suspect you belong in that group.

Naomi: I do.

John: Mehlberg says that for our purposes the substantivalist theory can also be called the "absolute theory." The key idea is that absolute motion is change of position with respect to substantive points of spacetime.

Naomi: So the substantivalist theory agrees that changing phenomena imply time but not that time implies changing phenomena. Is that right?

John: Yes, you've got it. Historically, Newton and Kant, the later Kant, are the substantivalists. Aristotle and Leibniz are the relationists.

Naomi: Does a substantivalist have to believe that time is infinite in the future and the past?

John: No. Some do and some don't. Many would say past time goes back only to creation, when God created both the first time and the first event. For these substantivalists, time isn't independent of God's actions. God can stop time whenever he wants.

Naomi: What happened to your idea that for any time there's an earlier time?

John: I guess there's an earlier time unless God decides differently.

Naomi: Well, I'd take that as evidence the idea isn't so a priori after all. Anyway, I wonder how you could measure the substantivalists' absolute time without something changing.

John: Everyone agrees time can't be *measured* without changes. Our issue is whether time *exists* without changes. It's a metaphysical issue, not an epistemological one.

Naomi: OK, but what does the substantival theory mean by change? When I think of change I usually think of having contrary properties at different times, like when a leaf is green all over and then later is partly brown. Is that the kind of change that's meant?

John: Yes, this ordinary kind of change is called "first-order change." It happens when an object in three dimensions varies its ordinary properties from one time to another. Second-order change is when first-order changes go farther and farther into the past, or get closer to the present from the future. When supporters of the substantivalist theory say there's time without change, they mean there's time without first-order change. They couldn't be ruling out second-order change.

Naomi: That's a helpful distinction. Maybe we also ought to try to get clear about what events are.

John: All right. I think of them this way. A song is a *kind* of event. The singing of a specific song by a specific person at a definite time and place is an event and not a kind of event. Specific changes, processes, and completed tasks are events.

Naomi: OK, that's a technical use of the word "event" because most people wouldn't distinguish an event from a kind of event. In physics we make another deviation from ordinary conversation and talk of point events even though there's no change involved. It's a helpful idealization that lets us use calculus for analyzing change. A specific leaf being green at noon at a specific place in Vermont is called a point event or instantaneous event even though there's no change in the leaf. Longer lasting events are composed of their instantaneous events. Are you willing to accept instantaneous events?

John: I'm not sure. They don't last long enough to be events in my usual sense. I'll decide about this later. For now, though, I'll go along with your idealization and say points events are events, too.

Naomi: Great. So were any reasons offered in your class for or against the substantivalist theory?

John: Here's an attack on the substantivalist theory. Try a little thought experiment of comparing our world—that is, our universe—with a world just like ours except that it began seven days later. These would be two different worlds for the substantivalist, but there'd be no discernible difference between the two. By Leibniz's principle of the identity of indiscernibles, the two really would be one. That's a contradiction because we started with two, not one. Therefore, the substantivalist theory is incorrect.

Naomi: I don't like the substantivalist theory, but I wouldn't want to reject it for this reason. I think substantivalists could defend their theory by saying times are properties of events in the world. The two worlds have different times and so different properties; therefore it's a mistake to say the two worlds are really one.

John: I agree with you. Thank you for defending the substantivalist theory.

Naomi: You're welcome. No charge.

John: It's interesting to think about Aristotle's attack on the substantivalist theory. His argument is that time without change is inconceivable. Therefore, we should accept relationism instead of substantivalism.

Naomi: *Is* it inconceivable?

John: No, and here's a thought experiment from the 1960s that convinced many metaphysicians they could conceive of time without change. Imagine all of space divided into three nonoverlapping regions. Assume each region undergoes freezing and unfreezing at a different rate so that
  Region 3 freezes every third year for a year.
  Region 4 freezes every fourth year for a year.
  Region 5 freezes every fifth year for a year.
There's no Region 1 or Region 2.

Naomi: Brrr.

John: Freezing doesn't mean everything's covered with ice; it means there's no first-order change.

Naomi: I know.

John: OK. Communication between regions allows those who have been frozen to learn after the thaw that time marched on in their neighbors' unfrozen regions while they themselves were frozen. Then their recently thawed clocks are reset to be in synch with their neighbors' clocks that weren't frozen. Now, here's what happens. Region 3 is frozen in years 3, 6, 9, 12, and so on. Region 4 is frozen in years 4, 8, 12, and so on. See how this freezes Regions 3 and 4 during year 12? When they unfreeze at the end of year 12, their neighbors in Region 5 will tell them to add a year to their clocks and calendars. After some experience with all this freezing and thawing, everybody in all the regions gets used to the idea that they themselves periodically freeze but then unfreeze. Now look at year 60. During year 60, every region freezes because $60 = 3 \times 4 \times 5$. At the beginning of year 61, they all add a year to their clocks, but otherwise there's no news to catch up on over that past year. Very boring during year 60.

Naomi: They can even predict when the universal freeze is coming. I'm convinced.

John: Great. What are you convinced of?

Naomi: Of the *conceivability* of time without change, despite Aristotle's claiming it's inconceivable. But think about the scenario with the three regions freezing. I had a thought about year 61. What could cause the universe to thaw at the beginning of year 61? It can't be something happening a minute beforehand because nothing is happening then.

John: I never thought of that.

Naomi: Well, I'm just wondering about the cause of a thaw. Now forget about science fiction and think about the real world. Why would any real scientist ever suppose there's empty time—time flowing when nothing is happening? I can't imagine any real, useful scientific theory requiring it, and if I can undermine the need for empty time, then I've provided support for relationism.

John: Yes, but there are reasons you need empty time. We all have this intuition that time marches on whether we live or die. It marches on regardless of what does or doesn't happen, so it marches on even when nothing happens. But if that's right, then we need substantivalist time, not relational time.

Naomi: I don't think an intuition that I'm wrong is an argument that I'm wrong.

John: OK, there are other reasons. Let's accept the big bang as the universe's first event. Isn't it intuitively clear, and logically consistent, that time could have existed for a long time before that, maybe even an infinite time, and then 13.7 billion years ago the universe began? So something could have happened before time began. See the contradiction? That means there's something wrong with relationism.

Naomi: I believe there *isn't* time without change, but I'm not sure I'd want to say there *can't* be time without change or that I can't conceive of it. I can conceive of a cow living on the surface of the sun, and I don't have a problem with saying this is logically possible. But it's not physically possible because real cows aren't like that; physical principles imply the cows can't take the heat. Similarly, I can conceive of time before the first event, and I don't have a problem with saying the substantivalist theory is logically possible, but it's not physically possible because real time isn't like that. Time is really dependent on events. I think you have this mental image in which the big bang begins and is the first event, but, before that, God might be doing something—making preparations for the first event. That's a contradiction right there according to my idea of what the relational theory says. I think there's no God-time, just plain, ordinary time; and everybody who's anybody is in it, like it or not. The only time intervals that are real are the ones between one event and another, like the time interval between the big bang and my riding on this bus. Time is how events are related to one another. But I don't want to rule out sensible talk about logical possibilities such as how our big bang might have happened earlier. Maybe there's a logically possible world similar to ours in which the big bang occurred 47 billion years ago. There's no contradiction in that.

John: I don't understand why you say you know the substantivalist theory isn't physically possible.

Naomi: Because if the first event occurred spontaneously even though time existed a week before that, then this would contradict the law of conservation of matter and energy. There's a violation because the sum of matter plus energy all week before the first event would be zero, but suddenly the sum jumps to being very, very large as the first event occurs. The jump violates the law. That's why empty time is physically impossible.

John: So dump the law, or make it have an exception.

Naomi: The law deserves more respect. It's proved itself to be very helpful in all areas of physics. We'd need a very good reason to throw it out. If you're going to accept substantivalism because you don't believe the connection is very intimate between time and change, then I could ask you why the first event happened at the time it did. Why did it happen 13.7 billion years ago, and not earlier? You need to answer this question, but I don't because I don't allow there to be any time before the first event. Can you say why the first event happened when it did and not a week earlier?

John: Maybe, but do *you* know why the first event happened the way it did?

Naomi: That's a different question. You and I both need to answer that one if you mean why did the universe begin with the event it did begin with. But I don't need to answer the question of why the first event occurred *when* it did. You do.

John: OK, now I see the difference in the questions.

Naomi: One of the merits of my relational theory over your substantivalist theory is that it escapes the question of why the first event occurred when it did.

John: I don't see what's so wonderful about escaping questions. Let's just answer them. The universe started when it did because God made the decision to have it start then.

Naomi: Groan.

John: I have another good reason for the substantivalist theory. Ever since the discovery of spacetime back in Minkowski's time . . . When was that?

Naomi: 1908.

John: Yeah, in 1908 he discovered that space and time are intimately related. They're so intimately related that you couldn't have a relational theory of time with a substantivalist theory of space.

Naomi: OK, I'll buy that.

John: But I have good arguments for the substantivalist theory of space, so that supports my theory of time, too.

Naomi: I need to hear the arguments.

John: I got this argument from Isaac Newton, but he used a bucket. I'm going to use a carousel, a round platform where kids ride on wooden horses that go up and down on poles while the whole platform spins to awful music. Now imagine that a few feet away the carousel is surrounded by a concentric wall. And there's nothing outside the wall. No earth, no stars. The carousel spins, and the wall stays still. But compare that with a similar world except that the carousel remains still and the wall spins the other direction. On the substantivalist theory of space these two worlds are different, but to a relationist there's no discernible difference of relationships inside the worlds, so all your experiences sitting on a horse should be the same in the two worlds.

Naomi: OK, they're the same. You've just got one world described in two different ways.

John: But they're not the same. If the carousel is still and the wall spins, then you feel nothing sitting on your horse, but if the carousel spins and the wall is still, then you feel a centrifugal force. If the carousel spins fast enough, the force would throw you off your horse. The relational theory can't account for the feeling of centrifugal force, but the substantivalist theory can. Centrifugal force is acceleration relative to substantivalist space. That's why there's a difference. But if that's the best explanation, then it's right to say there really is substantivalist space. This argument is Newton's. Leibniz and the other relationists had no comeback.

Naomi: I have a problem with the assumption that I'd be thrown off the horse if the carousel sped up. You're assuming this because of your own experience with spinning in our world with earth and the planets and stars out beyond every carousel. Is it so obvious I'd be thrown off if you removed everything beyond the wall?

John: Yes.

Naomi: Not to Einstein. He said, without the stars out there, spinning the carousel wouldn't produce any centrifugal force. Now get this. He said that if you were sitting quietly on a stationary carousel and then the stars were suddenly to spin rapidly around you, this would push you off the carousel in a tangent. I know this is unintuitive, but that's how it is according to general relativity. So you don't need substantivalist space, and Leibniz can rest in peace.

John: Hmm. OK, let me try another argument for the substantivalist theory. Consider a pair of perfectly matched hands, except one is left and one is right. Think of a world containing only the left hand, and another world containing only the right hand. On the relational theory there'd be no difference in the two worlds—no difference in how one part of the hand is related to another part—so the relationist would have to say there's just one world here that has been given two descriptions, but we know intuitively that the two are different. Left hands simply aren't right hands. The substantivalist theory can account for this intuition by saying the two hands have a different orientation with respect to substantivalist space. Therefore, there's such a thing as substantivalist space, and the relational theory is incorrect.

Naomi: Hey. You're pretty smart to figure that out.

John: I'm not smart. I just imagined some guy smarter than me; then I asked myself what he'd say.

Naomi: Ho. Cute. Who was it really?

John: Kant.

Naomi: Oh. Well, all I can think of to say is that maybe I'm wrong to believe that there's really a difference between a left hand and a right hand once you remove everything else from the universe. Look, I'll agree that you do have some reasons that favor the substantivalist theory, but I've always heard that Einstein's theory of relativity refuted absolute spacetime.

John: Mehlberg says you have to be careful saying that. There are two senses of "absolute." As we've been using the term, it means independent of the events, but as you used it just now it may mean independent of reference frame. Einstein's theory does show that duration depends on the reference frame chosen for measuring the duration, but this by itself doesn't show duration is dependent on events. The early Einstein did believe it showed this, but we have to be careful to distinguish Einstein from Einstein's theory. Einstein was very popular in the twentieth century, especially after ghostwriting the Rolling Stones song "Time Is on My Side."

Naomi: That's right.

John: Anyway, his prestige encouraged scientists and philosophers to agree with all his thoughts about the relational theory. It's just like when the philosopher John Locke read the scientific works of Newton and then was influenced to accept Newton's substantivalism. In the 1920s, when the philosopher Hans Reichenbach read the works of Einstein, he was converted to Einstein's relationism and tried to tell everybody substantivalism was dead. But I don't think it's so clear that Einstein's relationism can be extracted from Einstein's physics. Many philosophers are now arguing that you can accept both relativity and substantivalist spacetime while rejecting Newton's three-dimensional ether. Also, even if time were dependent on reference frame, it doesn't follow that spacetime itself is dependent on reference frame; maybe all that's dependent on the reference frame is how spacetime divides up into its time part. Then spacetime could exist all by itself without events. It would be an independent substance.

Naomi: I've never heard of any scientist who knew about relativity theory and accepted this.

John: There are a few out there, according to what I've heard. Here's why I think substantivism and relativity theory are consistent. Time moves from noon to midnight in your closet even when the closet is empty, doesn't it?

Naomi: My closet is never empty, but I'm proud of my closet. I have a pet name for it: Victory over Chaos. . . . OK, so time exists in empty closets.

John: Yes, and it's because events could have occurred there then but didn't, don't you think?

Naomi: Yes, but I'd emphasize the word "could." I don't think I can make sense of this with just the actual events.

John: You're going to need to make sense of points of space and time where the events *could* have occurred. So, even if you try to hold onto the relational theory, you're going to need spacetime points to be permanent possibilities of the location of events, and that's what the substantivalist theory has been saying all along, or at least a substantivalist theory that considers spacetime points to be substantial and not just God's thoughts or something. A spacetime point is nothing but a place and time where something *could* happen. So we can have our substantivalist theory and our theory of relativity. They aren't inconsistent.

Naomi: There ought to be some way out for a relationist. Maybe I could say each point of spacetime is the set of events that might occupy it in some possible situation or other, so it's events that are basic, not points.

John: You're going to accept possible worlds?

Naomi: You're right. I don't want them, but I was using them.

John: Look, there are other reasons to accept the substantivalist theory. I don't know the details of relativity theory, but I'll bet it has statements like, "For all spatial points x and times t, such and such an equation about x and t holds," right?

Naomi: Sure.

John: So relativity theory is committed to the existence of those things that are the x's and t's, but these are just what the substantivalist theory is saying exist. There you are.

Naomi: I'd like to treat points as-if rather than literally, or else explain them away as patterns of objects and events, but I can't see how to develop either idea right now.

John: If you can't, then I've made my case that there's such a thing as change of position with respect to substantive points of spacetime, and this is what absolute motion is. And there's another very good reason for substantivalist spacetime points. You can't do physics without fields, as I learned in your class. Gravitational fields and electromagnetic fields and other fields cover all of space. McClellan said the nonfield way to think of gravity in classical mechanics is with any one particle acting instantaneously on another particle. "Action at a distance," he called it. But he said the field way is better and it says there'd be this specific value for the gravitational force on a little particle if it were here, another value on the particle if it were there, and so forth over all of space. The field is the distribution over space of the force. Just like gravitational fields, the electromagnetic fields are distributions of electromagnetic force over space. He said waves of energy propagate through the electromagnetic field as the field values change with time, and that's really what light waves are. You're the physicist, so you know more about these fields than I do. Anyway, think about all this metaphysically. What are fields and what's waving? These fields can't be states of some Newtonian ether that waves when light waves if Einstein was right about that. But there must be something to have the field values at times. What else except spacetime points? We have to attribute the field quantities to substantivalist spacetime points. In other words, spacetime points are entities in their own right.

Naomi: [groans] That's a good argument, too. And I can't figure out how to say a physical vacuum has a geometry without believing in those physical points of yours. I thought relationism was the champ, but now I see why substantivalism is still a contender.

John: Great. Tell some of the other physicists.

Naomi: I will. Here comes your stop.

John: See you next week. They're predicting the first rain then.

Naomi: Don't meteorologists have a special inability to see farther into the future than two or three days?

John: Yeah, the fog of chaos is too thick. But every year it rains this month. Bye.

DISCUSSION QUESTIONS

1. Why is there something rather than nothing? If someone disagreed with you about this, what sort of reasons would they be likely to offer?
2. Did time have a beginning? What was that like? How can we know this if no one was there to witness it?
3. Which came first, the laws of nature or the first event?
4. Describe and discuss Kant's contradictions (he calls them "antinomies") about time.
5. Discuss this reasoning by Antonio Baberini: No matter how well a scientific theory of time or of anything else explains the effects we observe, one cannot know that the theory is true, since an omnipotent God has the power to produce the effects in some other way.
6. (a) Assess the explanation that says our specific world came into existence because all possible worlds exist. Defend your assessment. (b) Assess the explanation that the universe had to exist or else we wouldn't be able to be here asking the question of why it exists. Defend your assessment.
7. What's the difference between first-order change and second-order change? Can there be one without the other?
8. Why does John need to answer the question of why the first event happened when it did, but Naomi doesn't need to? What was John's answer?
9. Summarize the reasons mentioned by Naomi and John for substantivalism and relationism.
10. Are spacetime points necessary or contingent? Why?

FURTHER READING

Aristotle, *Physics*.
   Book IV contains the first relational theory of time.
Choi, Charles. "New Beginnings." *Scientific American* (October 2007): 26–29.
   A one-page discussion of whether physicists should believe there were events before the big bang.

Davies, Paul. "Facing the Unknowable." In *The Edge of Infinity*, 130–45. New York: Simon and Schuster, Inc., 1981.

An informal discussion of the belief that every event that occurs in the universe has its origin in some preceding causes.

Gott, J. Richard. *Time Travel in Einstein's Universe*. New York: Houghton Mifflin Company, 2001.

This physicist, along with his colleague Li-Xin Li, originated the idea of how the universe can create itself. See pages 186–92 for an elementary explanation of their idea. The graphic on page 188 shows how it's done. The first detailed theory of how the universe could have come from nothing was proposed in 1969 by Edward Tryon.

Hawking, Stephen. *A Brief History of Time: The Updated and Expanded Tenth Anniversary Edition*. New York: Bantam Books, 1998.

In chapter 8, this theoretical physicist provides a popular scientific introduction to the topic of time and the universe. Hawking suggests that perhaps our universe originally had four space dimensions and no time dimension, and time came into existence when one of the space dimensions evolved into a time dimension. He calls this space dimension "imaginary time."

Kant, Immanuel. "Concerning the Ultimate Ground of the Differentiation of Directions in Space." 1768. In *Theoretical Philosophy, 1755–1770*, vol. I in *The Cambridge Edition of the Works of Immanuel Kant*, translated and edited by David Walford in collaboration with Ralf Meerbote, 365–72. Cambridge: Cambridge University Press, 1992.

Kant uses a left hand and a right hand to argue for substantivalist space.

———. *The Critique of Pure Reason*. 2d ed. Translated by Norman Kemp-Smith, 1787. www.hkbu.edu.hk/~ppp/cpr/antin.html.

See "The Transcendental Dialectic, Book II, Chapter II, The Antinomy of Pure Reason," for Kant's views on the beginning and ending of time. Time, for Kant, is not a property of things in themselves, nor a property of our consciousness. It is a necessary presupposition of, or representation of, our experience.

Leibniz, Gottfried W. "Time Is a Relation." In *Time*, edited by Jonathan Westphal and Carl Levenson, 44–51. Indianapolis, IN: Hackett Publishing Company, Inc., 1993.

Excerpts from Leibniz's attack on Newton's substantivalist theory of space and time.

Newton-Smith, W. H. *The Structure of Time*. London: Routledge & Kegan Paul Books Ltd., 1980.

Discusses a wide variety of arguments, including Kant's antinomies, regarding whether the past and future have an infinite duration.

Quine, W. V. O. *From a Logical Point of View*. 2d ed. New York: Harper and Row, 1963.

Chapter 1 ("On What There Is") argues that a formal theory is committed to the existence of objects that are quantified over in its true statements.

Shoemaker, Sydney. "Time without Change." *Journal of Philosophy* 66 (1969): 363–81.

The original thought experiment using differential freezing of regions of the universe to show the conceivability of time without change.

Sober, Elliott. *Core Questions in Philosophy*. 2d ed. Englewood Cliffs, NJ: Prentice-Hall, Inc., 1995.

Lecture 7 ("Can Science Explain Everything?" pages 73–79) discusses the issue of whether something can come from nothing. Written for beginning philosophy students.

Teller, Paul. "Spacetime as a Physical Quantity." In *Kelvin's Baltimore Lectures and Modern Theoretical Physics*, edited by R. Kargon and P. Achinstein, 425–47. Cambridge, MA: MIT Press, 1987.

Pages 430–31 contain a summary of Hartry Field's sophisticated argument in *Science without Numbers* that field theory requires substantivalist spacetime points.

6

McTaggart, Tensed Facts, and Time's Flow

Naomi: Hi, John.

John: Naomi, have a seat. Hey, that's Melissa behind you. Hello, Melissa, good to see you. Sit down with us.

Melissa: OK.

John: How come you're on planet Bus today?

Melissa: I'm going across the city to meet some friends and play soccer. They couldn't pick me up, but they'll drive me home.

John: [to Naomi] Melissa is a grad student in my department.

Naomi: Hi, Melissa. Nice to meet you.

Melissa: You, too.

Naomi: Hey, John, what's the music today?

John: "I Can't Get No Satisfaction" by the Rolling Stones.

Naomi: Your taste is all over the place.

John: I pretty much like everything: rock, classical, country and western, blues, jazz . . . well, maybe not polkas. Did you guys get wet today?

Naomi: Not too bad; there's a covered walkway from the physics building to the library.

Melissa: I had an umbrella.

John: Good. This morning during the rain I saw two elderly ladies waiting at the bus stop near the Figueroa Hotel with plastic bags over their perms.

Naomi: I've seen them there before; they always make me smile. But we really need the rain.

John: The woman on TV last night promised there'd be no rain all month. She said she'd resign if it rained.

Naomi: That's a lot to promise! What channel?

John: The National Geographic Channel. I was watching a documentary on reptiles in the Sahara Desert.

Melissa: [groans]

John: Ha, ha, sorry.

Naomi: How was your class today? Learn anything new?

John: Definitely! Today was McTaggart day. He was an early twentieth-century metaphysician who was one-third Scottish.

Melissa: I didn't know that. Wait, how can you be one-third?

John: You're right. Anyway, Mehlberg spoke up more than usual today, and I think he didn't like the student's interpretation of McTaggart's famous argument about time.

Naomi: What's the argument?

John: McTaggart says our concept of time is expressed in terms of two competing series of events, the B-series and the A-series. They're the same events, but described differently. The B-series is about what events happen before what other events, and McTaggart says the series doesn't capture a fundamental feature—that time changes. The A-series is about which events are past, present, and future; McTaggart says it does show that time changes, which is a plus, but the A-series is self-contradictory, so our concept of time is faulty, and if this is time, then time doesn't exist.

Naomi: Wow! That's a radical conclusion. My gut reaction is that even if his complaints with the two series were correct, we shouldn't conclude that time is unreal, but only that human thinking about time is inconsistent.

Melissa: Yeah, I agree with you.

John: Me, too. Nobody these days follows McTaggart all the way to that conclusion, but there's continuing disagreement about where he went wrong.

Naomi: OK. Can you tell me more about his two series?

John: First, McTaggart assumes the future is real, and time is linear, and there's no time without events. Then he says there are two ways of creating the same ordering of all events in time.

Sentences of the B-series, the B-sentences, are about events happening before some events and happening after other events and being simultaneous with still other events. A typical B-sentence is "Abraham Lincoln's birth happens before his assassination." Another is "Abraham Lincoln's birth happens more than ten years before his assassination." On the other hand, the sentences of the A-series, the A-sentences, are about being present, being past, being future, and qualities saying by how much. A typical A-statement is "Lincoln's birth is over a century in the past." The A-series is the one McTaggart thinks captures the essential dynamic nature of time because the whole partition of time into distant past, recent past, now, future, and so forth, is sliding along the sequence of events.

Naomi: So there's just a single sequence of events, but he's describing it in an A-way from past, to present, to future, and also in a B-way from earlier to later and worrying about which description is best?

John: That's it.

Melissa: Being a philosopher of language, I think the two series are easier to understand if we focus on language about the series and notice whether sentences about the series change their truth-values.

John: Mehlberg said something like that.

Naomi: What do you mean?

Melissa: Think of true A-sentences being made true by A-facts. The A-sentence "We're now on a bus in L.A." is made true by facts about the present. But yesterday the A-sentence wasn't true. The sentence changes its truth-values as time flows along, McTaggart would say. And think about the corresponding fact. The fact that we're now on a bus in L.A. wasn't a fact yesterday. Facts come in and out of existence; they're transitory or "tensed." The change in truth-values of A-sentences and the transitory character of A-facts are what the dynamic nature of time is all about. Compare that with the B-series and its B-sentences that are made true by B-facts. A typical B-sentence is "The event of Lincoln's assassination happens before the event of our being on a bus in L.A." This sentence doesn't ever change its truth-value; it's eternal. And the corresponding fact is not transitory. This feature of B-sentences is what gives the B-series its static character when compared to the A-series.

Naomi: I see the difference in the two series.

Melissa: If you're the kind of person who believes that the concepts of past, present, and future are essential to any adequate metaphysical account of time, then you promote the A-series relative to the B-series, and you favor what is called the "tensed theory of time." If, instead, you think the concepts of past, present, and future are subjective and not needed for an adequate account of time, then you promote the B-series and are said to favor the "tenseless theory of time." According to the B-theory, if there were no subjectivity, or no consciousness, then there'd be no past, present, or future. So there are really two levels of disagreement; the level of the world and the level of language about the world. The static-dynamic dispute is about the nature of time itself, but the disagreement over whether to accept a tenseless or a tensed theory is over the terminology that's needed to talk correctly about time. People who accept the static theory usually accept the tenseless theory, and people who accept the dynamic theory usually accept the tensed theory. So philosophers of time divide up into the B-people and the A-people.

Naomi: I think I'm with the B-people.

John: I'm not surprised. I looked for an A on your forehead, and it wasn't there, so that's when I became suspicious that you're the zombie.

Naomi: Cute! But I don't see anything special about this A-series. It makes me nervous somehow. I can't exactly say why. I don't like the idea of the present running along the time line toward the future.

Melissa: Well, one reason you may be nervous is that if A-sentences stopped changing their truth-values, then our world wouldn't have any change in it.

Naomi: Maybe that's what makes me nervous. Change is a lot simpler than McTaggart makes it out to be. If I say, "The leaf was green on Monday but is brown on Friday," isn't that a B-sentence that's eternally true?

John: Sure, if you specify dates for Monday and Friday.

Naomi: But it describes the leaf changing color. Doesn't McTaggart understand this?

John: He understands it, but he doesn't agree with it. Without A-sentences changing their truth-values and A-facts coming in and out of existence time isn't changing, and so there aren't any times at all, he'd say.

Melissa: That's why Bertrand Russell never invited McTaggart to any parties.

Naomi: What parties?

Melissa: Russell was a fan of the B-theory, like you, and he disagreed with McTaggart. Russell and McTaggart were colleagues at Cambridge, and from what I know of Russell's personality and his four marriages, I'd guess Russell hosted a lot of parties and didn't invite McTaggart.

Naomi: Were they enemies?

Melissa: Well, I'm just fantasizing; I really have no idea whether they were friends.

John: They really were enemies. McTaggart was very active in getting the university to terminate Russell's teaching contract for being a vocal pacifist during World War I.

Naomi: Ouch! But I still don't understand why McTaggart keeps wanting time to change. When a leaf goes from green to brown, that's a change in the leaf, not a change in time.

John: Time's change is about second-order changes, not first-order changes.

Naomi: Oh. Remind me of the difference.

Melissa: First-order change is when objects gain or lose an intrinsic quality, like the leaf's having a green color. Second-order change is when the leaf's being green recedes farther into the past. For McTaggart, real change in time itself is when the partition of events into past, present, and future is altered. That alteration requires second-order changes. That's why McTaggart believes the B-theory can't handle change, the change in time itself. This idea is expressed linguistically or semantically by saying that second-order change involves A-sentences changing their truth-values. For example, the A-sentence "Lincoln was assassinated today" changes from true on that tragic day to false any day after that. Truth-values change as the now marches along the time line. Semantically, this feature of truth-values is time's change, its flow. The B-series isn't dynamic in this way. Sentences expressing B-relationships just sit there statically without ever changing their truth-values. The B-series doesn't capture the flow of events, the second-order changes.

John: I'm with McTaggart on all this.

Naomi: These are radically different approaches to what time is. What did he say about reference frames?

Melissa: He never mentioned them, as far as I know, but I can see what you're getting at. A change in reference frame would alter the partition of events into past, present, and future. Also, some of the B-sentences would change truth-values because a pair of events can be simultaneous in one frame but not simultaneous in another frame.

Naomi: McTaggart's view needs to be updated by saying every reference frame has its own A-series and B-series.

Melissa: That probably would work.

Naomi: I'm also bothered that McTaggart exaggerates how events change relative to me. What's special about me?

John: What do you mean?

Naomi: I mean the A-series is much more subjective than the B-series. If I utter the B-sentence, "Lincoln dies after Aristotle dies," then I'm not personally involved, but if I say Lincoln's assassination is past, that's implicitly a sentence about the "now" and about me because it implies that I'm now speaking after the assassination rather than before it. The A-sentences are too much about my consciousness. That's what bothers me about the A-theory.

Melissa: McTaggart would be happy about bringing in consciousness. He's a British variant on Hegel, who believed the physical world is ultimately an illusion, and reality is a collection of immaterial souls held together by ecstatic communion with each other.

Naomi: What?

Melissa: I know; don't ask me.

Naomi: I won't, but I find it metaphysically repulsive.

Melissa: I think Hegel set back the growth of European philosophy by a hundred years.

John: Hegel was one of the greatest nineteenth-century philosophers.

Naomi: How sad.

John: Well, let me put on a McTaggart hat for a moment and defend him. I think McTaggart is right that there's a contradiction in the A-series because the A-sentence, "We're now on the bus," is both true today and false yesterday, so it's both true and false, which is incoherent, right?

Naomi: No! That's so odd! Did McTaggart really say this?

Melissa: Yes.

Naomi: Well, being true and being false are contradictory, I agree. But the sentence "We're now on the bus" has those truth-values successively, not at the same time. It seems to me McTaggart's mistake is like saying the cow is in the barn today but not in the barn yesterday, so we have a contradiction. That's silly.

John: McTaggart was aware of this objection and had a sophisticated response. It's complicated, very complicated. Think about today's lunch. Today is Thursday, so you'd say that the lunch *is now present* on Thursday, *will be past* on Friday, and *was future* back on Wednesday, right?

Naomi: Yes, that's why there's no contradiction. The three sentences can all be true; they're consistent.

John: McTaggart was ready for your response that the three can all be true. You've introduced three new second-level, compound tenses: is-now-present, will-be-past, and was-future.

Naomi: Maybe I did, but it's acceptable English, isn't it?

John: Sure, but here's the new problem. Your sentences capture only some of the A-series facts. To express all the other facts like this you'll need six more compound tenses for a total of nine. Here, let me write them down in groups of three: was-past, was-present, was-future; plus is-now-past, is-now-present, is-now-future; plus will-be-past, will-be-present, and will-be-future. The three you used from the nine don't happen to contradict each other. But the others do, McTaggart said. For example, given the second-order changes that events undergo, these two sentences are both true:
Thursday's lunch *is-now-present.*
Thursday's lunch *was-present.*
See the contradiction?

Naomi: Wait! Yes, I may need those nine tenses, but these two sentences aren't true at the *same* time, only at different times. See? No contradiction.

John: Fine, it looks like you've escaped the contradiction, but McTaggart predicted that response, too. It takes you to third-level tenses. And you'll again try to escape the contradiction by saying the sentences aren't true at the same time. You'll have added three new tenses for each of the old nine, giving you twenty-seven third-level tenses, and McTaggart will say that some true sentences will contradict others, and so on.

Naomi: Yes, but at each level I can remove the contradiction, so the regress isn't vicious.

John: But at each level I can find a contradiction, so the regress *is* vicious.

Naomi: Hmm, maybe this does show the inconsistency of the A-series. I don't like the A-series, but I don't think you can show a priori that it's contradictory. I'm starting to get a brain cramp. Didn't Socrates say the examined life is too difficult to live?

John: Something like that. Anyway, I was just defending McTaggart temporarily. Actually I agree with you that the A-series isn't contradictory and that an event which is both past and future is past at one time but future at a different time. I still believe in the A-theory and its tensed theory of time, but for our peace of mind let's bracket tenses and focus directly on the flow of time. I say McTaggart is right about this flow. You don't, do you?

Naomi: No, and now I know another reason why the A-series bothers me. McTaggart says what's so great about the A-series is that it captures the idea that events change. But I say, "Who needs them to change?" First-order change is the only change worth paying attention to. Second-order change is a sure ticket toward confusion. It leads you to make the mistake of saying events change and time flows and that you need multi-level tenses and God knows what else. McTaggart's flow has to do with all this second-order stuff, but it's just a subjective feature of everyone's experience.

John: It's too hard for me to believe the flow isn't really there. Don't you feel time flow?

Naomi: Yes, I feel what you feel, but I see optical illusions, too, and I don't say they're real. I recognize that if I say time's passage in a particular direction is an illusion, then I have to account for why the illusion has such a hold over us. My best guess is that it's due to our being aware that our memories are accumulating over time. More memory is all there is to time's flow. What is objectively true is that we do get new memories. But the sense of flow is all about the subjectivity of awareness.

Melissa: Or maybe time's flow is about anticipations of experiences happening before memories of them.

Naomi: OK, maybe that, but whatever explains our feeling of flow, time isn't really flowing out there in the world beyond our mind and brain.

John: No, the passage isn't just subjective. I'm sure it's connected with objectively coming into existence, like when future events become determinate by becoming present events. Becoming unfuzzy, I like to say. This process propels the now forward. I wrote down this great quote from Santayana that gets it right: "The essence of nowness runs like fire along the fuse of time."

Naomi: A fire can go slower or faster down a fuse, but does the flow of time have a rate? I don't think there's a rate.

John: How about a rate of one second per second?

Naomi: The units in the numerator and denominator cancel out.

John: Look, even if we can't measure the rate of that change, we have to accept it. The A-series accepts it; the B-series doesn't.

Naomi: Even though events don't change in the B-series, the series itself contains information about all the objective changes, the first-order changes. You can look at the series and see that change occurs to an object by its having incompatible properties at different times. Time is about change, but time itself doesn't change. The cosmos doesn't become real or happen or flow or anything strange like that; it simply is. We have a deep disagreement here.

John: Yes, we do.

Melissa: Maybe I can step in here and make a suggestion. It can help to focus on language—on sentences *versus* statements and on what makes them true or false.

John: What are you getting at?

Melissa: Statements are what gets stated when we use sentences to say something true or false. So statements always have truth-values. Sentences don't. The sentence "It's now noon" isn't true or false, but uttering it at noon will be using it to make a true statement, and uttering it ten minutes later will be using it to make a false statement.

John: All right, there's a difference between a statement and a sentence.

Melissa: And I'd like to recommend that when we use the word "statement" we mean something that has a truth-value that doesn't change with time. The technical terminology is that I'm talking about "eternal" statements.

John: I'm getting an idea of where you're going with this.

Melissa: Now think about what a fact really is. Facts are actual arrangements of the objects in the world. When we make a true statement by uttering a sentence, we state a fact. Loosely speaking, truths correspond to the facts. It's unclear whether we can make perfectly good sense of the notions of correspondence or of facts, but let's not worry about all that right now and just accept this correspondence theory of truth. Are you with me so far?

John: Makes sense to me.

Naomi: Me, too.

Melissa: Let's check. Take another typical A-theory sentence. For example, when Julius utters the sentence, "President Abraham Lincoln is dead," is Julius making a true statement or a false one?

John: A true one.

Melissa: Not if Julius is Julius Caesar.

John: You tricked me!

Melissa: Yes. The sentence all by itself has no truth-value because something essential is missing. What's missing is the context, the background information about where and when and who and all that.

John: OK, maybe. But isn't it a fact that Lincoln is dead compared to our present time?

Melissa: Yes, definitely. But my point is that, for this comparison, you needed to appeal to information contained in the context. All by itself, there's no fact picked out, no definite eternal statement made, see? You should say—well, what I'm recommending is—the sentence all by itself with an unspecified context has no truth-value because it's ambiguous as to which *statement* is being made, which *eternal* statement. But look what follows from this. If this typical A-theory sentence has no truth-value, then you shouldn't say it *changes* its truth-value. And if it doesn't change its truth-value, then you can't explain the flow of time as the change of the sentence's truth-value. Yet that's what McTaggart is trying to do with his A-series.

John: OK, that's what he's doing, but can't we say the sentence changes its truth-value as the context changes?

Melissa: Yes, we sometimes talk that way, but it's misleading because it suggests that the sentence itself has the truth-value. Strictly speaking, the only kind of sentences that could have truth-values all by themselves are context-explicit sentences such as "It's raining in L.A. on such and such a date" or else context-free sentences such as "2 + 2 = 5." Context-explicit sentences and context-free sentences are associated with just one statement, one eternal statement. It's clear what statement would be made with these special sentences because their truth-values don't change with time. Now think about collecting all the facts you need in order to produce an ordering of the events in one reference frame. To accurately specify the occurrence of an event with a sentence, you need to make the context explicit. To do this, you need to state the antecedents of pronouns, and insert the place and time of the event. If the event is an action, then you need to state the actor. You can't say going to the supermarket is an event. It's a *kind* of event. The event would be something better described as Leslie Williams's going to the supermarket on Fillmore Street at 5 PM on such and such a date. Once you accurately specify an event, you'll see that the event doesn't change from being present to being future, or change from being past to being farther in the past; it just exists. There's no flow. Good descriptions of events should be time-stamped and otherwise situated in a context.

Naomi: Yes, that sounds good to me; it's what I should have said.

John: I don't think it's fair of you to stipulate that most of the declarative sentences ever spoken fail to have truth-values.

Melissa: I think it's OK to revise definitions in order to clarify a problem or solve it.

John: Let me just make sure I understand. Take the sentence, "It's raining in L.A." I would say the statement made with this sentence has a truth-value that varies with time, and you would disagree and say the sentence has no truth-value because the time we utter the sentence is part of the context that determines what statement is made, thinking of statements as eternal statements.

Melissa: That's right. Naomi, where are you on all this?

Naomi: I tend agree with you, not John, but I haven't thought as much about the nature of language as you have, so I probably don't believe as strongly as you do.

Melissa: There's still the issue of tensed facts. They're supposedly what makes tensed statements be true or be false. But I think tensed talk shouldn't be taken too seriously or literally even when we do specify the context and get clear about what statement is being made and find that the statement has a tense. This is because talk using tenses usually can be translated into talk not using tenses. That's a plus for the B-theory and a minus for the A-theory. Here's what I mean. When you utter the A-series sentence, "Lincoln died in the past," the sentence can be translated by the philosophical analyst into the B-series sentence, "Lincoln dies before time t, and time t is the time of Melissa's uttering the sentence 'Lincoln died in the past.'" Look at the tense in what I just said. When I said "Lincoln dies before time t," the verb "dies" was really tenseless there. By "tenseless," I mean logically tenseless, not grammatically tenseless. All verbs have some grammatical tense or other. The word "dies" is in the present tense, but logically it's tense-free just the way "is" is tense-free in "The cow is a mammal." It's intended as "is eternally" or "is at all times." I didn't intend to say the cow is a mammal *presently*, and I didn't intend to say Lincoln dies presently. The "t" is specified tenselessly, too, with a date. But look at what we've got here. Tenses can be analyzed away. So we don't need any tensed facts to make statements be true.

Naomi: That seems right to me, a tenseless theory of time.

Melissa: And think about "now," the word. It's true that it's now past noon, but we don't need present tensed facts to make it true; all we need is to be aware that the reference keeps shifting around as time goes by. The B-theory appreciates that the word "now" is merely an indexical term with a slippery reference, so it doesn't need to treat presentness as flowing along the time-line fuse.

John: You're getting rid of the present!

Melissa: Yes and no. I'm trying to give it the ontological standing it deserves, but I'm not saying ordinary talk about the present is nonsense, or anything radical like that. Some events occur in the present and some don't; there's a difference, of course. But the deep point is that tensed terminology, such as "is present," isn't revealing something ontologically basic, and it would be misleading to try to understand it as being about the presentness in the objects being talked about, as if it's some nonrelational, objective property that sticks to events making them present until they lose the property and become past.

Naomi: I get it now! All the real facts can be stated tenselessly by using full sentences—the eternal sentences—that aren't vague or ambiguous or indexical by being relative to speakers and times and places. That's why McTaggart's A-qualities aren't objective qualities of events, and that's why A-theory sentences don't really state facts about time, or at least not any facts over and above the facts expressible with B-sentences.

John: I don't know if that's good enough. You're trying to make the tense unimportant, but I think that the A-theory's tensed verbs have an ontological role to play and can't be eliminated and shouldn't be even if they could be. The truths about human perspective and our subjectivity are some of the most important facts there are. These truths are expressed with tenses. You shouldn't translate this out of the picture and still claim to be giving a complete description of what people say when they speak the truth.

Naomi: Why not? Melissa isn't claiming of any truth that it's not a truth, but the A-series really is illegitimately person-relative. As a physicist I like to think there's a person-independent world out there where time exists objectively, and for this reason the A-series can't be fundamental to time; only the B-series can be. Think of this analogy between space and time. I can have a complete description of spatial reality without knowing which places are here and which are there. A map is complete and works for all people, regardless of where they are. It's not person-relative. The mapmaker removes the subjectivity. Similarly, I can have a complete description of temporal reality without knowing which events are in your past or in Lincoln's past. All I need is the B-series relations to get the objective facts.

John: Tense is more important than you give it credit for. For example, don't you prefer that a particular painful experience be in your past rather than in your future?

Naomi: Sure.

John: So you're biased toward the future. An A-theorist can account for this bias by saying you care more about the future fact "I will be in pain" than the past fact "I was in pain," but the B-theorist can't account for the bias this way. Or any way. All they have is the tenseless "I am in pain at time t."

Naomi: Well . . .

John: Don't talk to me about your caring for the time when sentences are uttered.

Naomi: Hmm. I can see that this is a problem for the B-theory. There's more to fear of pain than caring about sentence times, so I'll promise to think hard about how to solve this problem. But it still seems to me that from the scientific perspective, one person's perspective shouldn't be any more important than anyone else's. I'm simply promising that anything objective somehow can be adequately handled in science by the B-theory.

Melissa: But it can't be. I know I'm moving over to John's side in this dispute. Sorry, Naomi.

Naomi: OK, pull the trigger.

Melissa: Think about this scenario with your B-theory. Suppose John Perry is pushing his shopping cart in a Palo Alto, California, supermarket, when he notices a trail of sugar ahead. Following the sugar trail he comes upon a trusted friend who tells him, "John Perry is leaking sugar from his shopping cart." Perry believes this, but he has amnesia and doesn't know he himself is Perry. In this unusual situation, how is the B-theory going to explain Perry's belief? It will only be able to say Perry believes Perry is leaking sugar from his cart. The B-theorist will attribute the same belief when there's no amnesia. So, the analysis leaves out something essential about the indexical, something essential about the vantage point, doesn't it?

Naomi: Yes, but I don't consider vantage points to be objective.

Melissa: Fine, they're subjective, but then your B-theory, and science in general, isn't doing its job because it's missing objective truths about our subjective experiences.

Naomi: I'm not sure. OK, I'll agree the A-series really does have something that the B-series doesn't. It expresses an important way that most people think about time from their own personal perspective. But this expresses something about people rather than about the ontology of time.

Melissa: Ontology should include people's realities, shouldn't it?

Naomi: Now I don't know what to say. Maybe you're right.

John: What? Did I just wake up and hear what I thought I heard? How long have I been asleep? What century is this?

Naomi: All right, all right. I don't always agree with myself.

Melissa: Naomi, maybe I can make another point here that helps you. Admit that we *can't* translate "Perry is leaking sugar" or "I'm now in L.A." into tenseless sentences without losing something important about vantage points. And admit they can't be translated adequately into indexical-free sentences. Fine. Then say this is just an interesting fact about our language, but say this doesn't tell us that there are tensed facts making tensed sentences true. Tenseless facts can account for the truth conditions of tensed sentences For example, take the A-sentence, "I'm now in L.A." The sentence is true because it corresponds to the fact of Melissa being in L.A. at time t, where t is November 5, 2009, at . . . uh . . . I forgot to wear my watch today. Simple. Tenseless facts accounting for the truth of a tensed sentence without having to find a tenseless translation for the tensed sentence. Victory for the B-theory.

Naomi: I can hear the trumpets. Thank you, Melissa!

John: That tactic won't always work. When my pain ends, I say "Thank goodness that's over" because I'm thankful for the pastness of the pain, for the fact that the pain is in the past and not in the present. See? We need this tensed fact—the fact that the pain has pastness—to explain why it's true. I'm calling for the B-theory to surrender.

Melissa: No. It's not the *pastness* of the painful event that explains why you say, "Thank goodness that's over." Your relief is explained instead by your *belief* that the event is past, plus its being true that the event occurs before the utterance of the sentence. There. No tensed fact is needed. Talking about the quality of pastness is misleading, I think. An event can be past without there being a quality of pastness that the event has, just as an event can exist without there being a quality of existence that the event has. So that's why we should forget about the A-theory.

Naomi: That sounds great!

John: You still have problems. You're relying on sentences being uttered. Aren't sentences true even if they're not uttered?

Melissa: OK, replace utterances with thoughts. Use the B-sentence that says the event occurs before the time of your thought, where the thought is expressible with that sentence. See, we have tenseless facts about thoughts explaining why your tensed talk is true even if it can't translate 100 percent of your A-talk into B-talk.

John: But can't a sentence be true even if no one has thought it?

Melissa: Yes, so let's make another revision and use possible thoughts instead of just thoughts. That works. In addition, I'll bet tenseless facts can be used to explain the logical relations between tensed sentences—that one tensed sentence implies another, is inconsistent with yet another, and so forth. Then I'd apply Occam's razor. If we can do without essentially tensed facts in accounting for tensed talk, then we should say essentially tensed facts don't exist.

Naomi: Hand me my razor, please.

John: I don't believe tenseless facts can account for every truth worth explaining, so I'd like to see a development of this idea before I let you two loose to slash away with Occam's razor. I wonder if you can distinguish a possible thought from an impossible one.

Melissa: That's a fair challenge. So where are we now? You and Naomi have very different metaphysical positions. You two have a static-dynamic dispute and also a tenseless-tensed dispute. Your static-dynamic dispute is over the nature of time itself and the role of change and whether time really flows. Your tenseless-tensed dispute is over what concepts are necessary for describing time, and it's over what kind of facts must exist out in the world to account for the truth of our sentences that use those concepts. These are all tough issues to settle.

John: Yeah, for sure. Hey, my stop. Melissa, thanks for clarifying some of those issues for us.

Melissa: Happy to take part.

John: Hope you have fun at your soccer game.

Melissa: Thanks. I will.

DISCUSSION QUESTIONS

1. Is the sentence, "John had lunch at noon today" expressed in A-series talk or B-series talk? Why? How about the statement that John had lunch at noon?
2. Is the A-series contradictory, or should we say the A-series is consistent but McTaggart's reasoning about it is not? Why?
3. Are B-series facts determined by more fundamental A-series facts, or is it the other way around, or are both suggestions in error? Begin by explaining what the question is asking.
4. Does time flow because events change? What is mistaken about your critic's reasoning?
5. John claims the B-theory misses truths about vantage points. What's his point?
6. What's the best way for the B-theory to try to express the fact of what time it is now?
7. What does it mean for something to be grammatically tensed but logically tenseless?
8. How should the B-theory try to explain the remark about the cessation of pain when we say, "Thank goodness that's over"? Can the B-theory succeed at this?
9. Explain the point Melissa is making with the scenario in which Perry is pushing his shopping cart in a Palo Alto supermarket and notices a trail of sugar ahead. How good is Naomi's response to Melissa?
10. Melissa believes there is an A-statement that can't be translated into a B-statement. Why? What's left out?

FURTHER READING

Dainton, Barry. *Time and Space*. Montreal: McGill-Queen's University Press, 2001. Pages 10–12 and chapters 2 and 3 contain helpful explanatory material on McTaggart's A-series and B-series.

Davies, Paul. *About Time: Einstein's Unfinished Revolution*. New York: Simon & Schuster, 1995. A physicist surveys the problem of time's flow in chapter 12.

Dummett, Michael. *Thought and Reality*. Oxford: Clarendon Press, 2006. Chapter 1 discusses the differences among facts, propositions, sentences, utterances, and statements. See pages 11–13 for a brief overview of the dispute between Prior and Frege on whether sentences have truth-values that vary with time. The rest of the book is difficult reading for those who haven't studied philosophical logic. Dummett is a justificationist and not a realist about truth because he believes a proposition about the past, future, or present is timelessly true if and only if "someone optimally placed in time and space could have, or could have had, compelling grounds for recognizing it as true" (viii).

Garrett, Brian. *What Is This Thing Called Metaphysics?* Abingdon, Oxon: Routledge, 2006.
Chapter 5, "Time: The Fundamental Issue," discusses McTaggart and the A-series and B-series at the same intellectual level as in our dialogue.

McTaggart, J. M. E. "The Unreality of Time." *Mind* 17 (1908): 457–74.
The original statement about the difficulties with the A-series and the B-series.

Perry, John. "The Problem of the Essential Indexical." *Nous* 13 (1979): 3–21.
An influential argument that indexical sentences cannot be reduced to indexical-free sentences.

Prior, Arthur N. "Thank Goodness That's Over." *Philosophy* 34 (1959): 17–23. Reprinted in *Papers in Logic and Ethics*, by Arthur N. Prior, edited by P. T. Geach and A. J. P. Kenny, 78–84. London: Gerald Duckworth & Company Limited, 1976.
Argues that a tenseless theory of time fails to account for our relief that painful past events are in the past rather than in the present.

Quine, W. V. O. "Time." In *Word and Object*, by W. V. O. Quine. Cambridge, MA: M.I.T. Press, 1960.
This short, five-page section provides Quine's influential argument for the tenseless theory of time and how, from a logical point of view, time should be treated as if it were an extra dimension of space, and objects should be treated as events.

Williams, Donald C. "The Myth of Passage." *The Journal of Philosophy* 48 (1951): 457–72.
Williams discusses the pros and cons of believing in time's flow and concludes that it doesn't flow.

7

Presentism, the Block Universe, and Perduring Objects

John: Ah, the photon collector has returned.

Naomi: Hi, John. You have a thoughtful face on today.

John: Well, my supervisor at the restaurant is making my life miserable right now. He keeps pressuring me to take the night shifts nobody else wants. I told him I'd think about it because he's one of those people you really can't say no to.

Naomi: Too bad. Maybe someone else will step up and take the shifts before things get too messy with your boss. Do you get to eat your meals there?

John: Yeah, it's one of the perks that comes with the job, but our house specialty sits on my stomach like a bagful of buckshot, so I do mostly soups and salads.

Naomi: Good choice. How's school?

John: That's a better part of my life. I'm loving my time seminar.

Naomi: What was on the menu today?

John: We covered the three big ontologies of time: presentism, the growing universe, and eternalism.

Naomi: Let's see if I remember these. Presentists don't like the Beatles' lyric "Oh, I believe in yesterday." You growing-universe people do believe in yesterday, but not tomorrow. Eternalists say the past, present, and future events are all real.

John: More or less. The growing-universe theory implies the past and present are real. Presentism implies reality is only the present. What do you think of presentism?

Naomi: Its reality is too brief. A bullet's motion is all about the bullet's position at times other than the present. Its speed is the rate of change of position from one instant to a nearby instant. This idea is expressed in calculus by saying speed at an instant is the derivative of position at that instant, but the derivative involves positions before or after the instant. If you focus just on the present instant, you'll never understand present motion; it'll be a mystery why it occurs at all.

John: I don't understand calculus, but I suppose the presentists would say you can treat motion *as if* it works the way calculus says, and they'd emphasize the "as if."

Naomi: Seems too cautious to me. How do you suppose presentists explain that Lincoln was assassinated in the past?

John: They'll use present traces of what you and I call the past assassination. The traces would be his grave and the historical documents in newspapers and that sort of thing. A student in my seminar showed how they'd do without past-tense statements like that. They'd change formal logic by allowing the past-tense operator "It was the case that . . ." to operate on the present-tense statement "Lincoln is assassinated." Then they'd say what makes this statement with the operator true aren't facts about the past but only facts about present traces. But here's the tricky part. They say the present traces actually *constitute* the fact that Lincoln is assassinated. I can't believe that. I mean, what if there were no traces left? Would the fact disappear? If presentism is committed to that, I'd say, "*Hasta la vista*, baby."

Naomi: Yeah. Also, aren't facts about the past supposed to *cause* facts about the present? If the presentist tries to tell me causes are unreal, then ditto on the "*Hasta la vista*."

John: In their favor, though, I can see why treating the past tense using operators on present-tense sentences makes sense; it captures how children learn about the past. Think how they learn about the meaning and truth of the sentence "Our house was cold." First the child learns what "Our house *is* cold" means, and the child learns how to judge whether it's true. Only later does the child apply that understanding to figure out what "Our house was cold yesterday" means and what it takes to settle on whether it's true. It's like the child learns how to operate on the sentence "Our house *is* cold" with a past-tense operator. Then they learn to extend those ideas about yesterday to days farther in the past.

Naomi: Yes, that's probably how everyone learns, but I don't see why metaphysical claims about whether the past is real should be influenced by how everyone learns.

John: Well, there's another big reason to be a presentist: the present is more tangible than the past. You can't touch yesterday's dinner right now.

Naomi: I agree, but that doesn't convince me to be a presentist. I believe in things I can't touch. Besides, I have an instrument in my physics building that can detect the past.

John: You do?

Naomi: It's a telescope. If you point it at the North Star, you see it the way it was 680 years ago because the star is 680 light years away from Earth. The star might not exist now; it might have burned out before Lincoln was assassinated. Here's an idea! Suppose there is a gigantic mirror out there by the North Star 680 light years away. We could stare into it with our telescope and see live action on Earth as it was thirteen centuries ago. Isn't that a good scientific reason to believe the past is real?

John: To me, yes. What would a presentist say? Probably that when you look into a telescope you infer the past but don't see it. Something like that.

Naomi: A desperate remark.

John: Yeah. For me, what makes something real is that it's no longer just a possibility. Reality grows with the coming into being of determinate reality from a merely potential reality. Reality grows by adding new facts. It's all about becoming real.

Naomi: The past grows by adding facts, sure, but I don't think the past is real for that reason. I look at reality differently. Time isn't space, but a time dimension is a lot like a space dimension. I think of time as a single dimension of a long block of four-dimensional spacetime. Well, it would be a block if none of the dimensions were infinite. A plane through the block at the time I'm speaking to you is the plane of the present if it's perpendicular to the time axis. The plane extends throughout 3-D space. Technically it's called a "hyperplane" instead of a plane because it's three-dimensional and not two-dimensional, but in a Minkowski diagram with two space dimensions and one time dimension, it's an ordinary plane. The plane separates the future from the past. Yet it's merely a person's personal perspective of when now is occurring that fixes this plane. The whole block is real no matter how some observer's mind slices it up into a future part and a past part.

John: I understand your block metaphor, but I think it's misleading. First, it makes the objective distinction between future and past seem to be subjective. Second, it gives time a very static treatment. There's no flow of time. The block just lies there—like the bagful of buckshot.

Naomi: True, but for me, one advantage of the block universe is that it avoids the weirdness about time's flow.

John: I'd turn that remark on its head and say the block theory can't explain time's flow, so it's a faulty theory of time. And what makes you so sure the future should be a part of the block? It doesn't seem very real to me. It's not definite or settled. Show me something affected by the future.

Naomi: Being able to affect the present or past isn't a requirement I'd place on being real. By the way, by saying "future events," I mean actual future events, not merely possible ones. The human predicament is that we don't know which possible future events are actual future events. But because the actual future is real, it can't be changed.

John: I think the future *can* be changed. That's one reason why it's not real. Are you implying future events are sitting out there in the future waiting to happen?

Naomi: Sort of. It's there for the visiting with a time machine, just the way the past is. Look, you're going to die someday, right? Your death hasn't happened because it's off in the future. So there's a future it's off in.

John: You're arguing like this. He's off hunting ghosts in the attic, so there are ghosts he's off hunting.

Naomi: Oops! Note to self: Turn on brain before opening mouth. That wasn't such a good argument. What I mean is this. Back in 1700, was the 1865 assassination of Lincoln by Booth sitting out there in the future waiting to happen, no matter whether the assassin is born? No. If Booth had never been born, then the assassination by Booth wouldn't have occurred. Anyway, the point is that future events that do happen are usually very dependent on what happens earlier. Change those earlier events, and nature usually will produce a different future.

John: That makes you a causal determinist, someone who believes there are laws of nature that operate on how things are at any one time to determine how things will turn out at later times.

Naomi: I'm no causal determinist.

John: I think you have to be if you buy into the block universe theory. The way I look at it, our future a thousand years ahead already exists right now for God, and not for us. The future is "open" for us; there are so many possibilities. The past isn't like that. The past isn't open. That's why it's real and the future isn't. You don't think of the future as being open.

Naomi: Are you saying our future is real for God but not for us? That makes it real and not real!

John: Yes. I'm not afraid to say this. Look, there's another perspective for discussing this whole issue. Our disagreement about the ontological character of the future can be turned into a disagreement about the truth-values of sentences. Remember our earlier discussion about Aristotle saying sentences about the future sea battle aren't either true or false now? I'd say the sentences have no truth-values at the moment because the future is unreal; right now there are no facts making the sentences true or false. It's only later that they acquire their truth-values.

Naomi: You're all worried that there's nothing now to make sentences about the sea battle be true, but who needs anything now? The sentences are about the future, so what makes them true or false is the character of the future, not the present. You're putting too much emphasis on the present and being too suspicious of future possibilities. You think that what's wrong with the future is that it's merely possible, but you may be mixing up two senses of the term "possible." It's possible that Abraham Lincoln drove a car to the White House, but that's not enough to make it actual. It's merely possible and so is unreal. OK, I'd agree with you on that. But your death or any other event in the future that will happen is different. It isn't *merely* possible. It's both possible and going to be actual. That's why it's real.

John: No, no, no.

Naomi: Yes, think of it from another angle. I'll bet you think abortion is usually immoral, and you think this partially because you don't want a present fetus not to be a future adult. You care about that future adult. But why care about the future adult unless you think it's real? You care about what's real, not what's unreal, don't you? So you're treating the future with this adult in it as being real. Even *you* accept the future.

John: No, I don't care about the future adult. I care about present immorality!

Naomi: Relax. OK, that angle didn't work on you. Let me try again. We all care about children having a good future. We want future generations not to choke on smog and have to drink polluted water, right? So we're making moral judgments about future people and future states of the world. We're treating them as being real because they *are* real.

John: No, they're just treated *as if* they're real.

Naomi: You're a tough customer. Maybe you'd accept the reality of the future if you were to discover a traveler from the future. Here's how it might work. Suppose some future civilization learns how to arrive here on Earth from the year 3005. This would be empirical evidence that future events exist ahead in the year 3005, events such as a traveler climbing into a time machine then. This wouldn't just be evidence that the events *will* exist. In other words, we'd have empirical evidence that eternalism is correct.

John: Hmm. Let me think for a second. Maybe all you've got here is an argument that time traveling from 3005 to now is impossible.

Naomi: You've turned my argument around on me.

John: That's what I was trying to do.

Naomi: OK, I can think of an even better reason to be an eternalist.

John: You can?

Naomi: The special theory of relativity implies eternalism. Isn't it reasonable to say that, if there's some reference frame in which two events are simultaneous and if one of the two is real, then so is the other? Like your talking to me now is real, and it's simultaneous with some event—say, an explosion—occurring in a faraway galaxy, so the explosion is real, too.

John: I suppose so.

Naomi: Well, relativity theory implies the past-future distinction is relative to the reference frame, and it implies there are no privileged reference frames. Judgments made in reference frames moving relative to each other will disagree on which events are in the future. That distant explosion can be in the present of your frame and in the future in my frame. Are you going to force me to call it unreal because it's in my future even though it's real for you? That's a paradox. The only way out is to agree that all future events are real for all persons. That makes the case for eternalism. I got this argument from reading Hilary Putnam.

John: I've heard of him. OK, I agree with you that it would be too odd to say that whether an event exists depends on how fast you're moving. But how does relativity theory handle the present?

Naomi: Having a single present for all of us is useful, so by convention we choose one person's frame as the privileged frame, such as the frame of the person who is at a place where the background microwave radiation is the same in all direction. There's not really a person there, but astronomers find it convenient to use this frame and to say two events are simultaneous if they exist in the same hyperplane in that frame. But it's all conventional; nobody is tricked into believing this is the "real" present in some deeper sense.

John: I'd like to accept relativity theory if only because the scientists say it's so useful, but it seems to me that what relativity theory implies is not eternalism but only that we can never physically find out which events are really present instead of really future, so we search around for some conventional present that we can all accept for doing science. I accept relativity theory up to a point, but it's incomplete because it can't tell which reference frame is the absolute one, the one that tells us how long any event really lasts.

Naomi: I don't think any improvement on relativity theory will find that frame.

John: If you're right, then maybe we're probing a level of reality that is beyond scientific methods. I still believe in the tensed theory of time that we talked about last week, so I still want to believe the presentness of an event is no conventional feature of nature, and no subjective feature. It's a simple, nonrelational, objective property of the event, and one it loses when the event is no longer present.

Naomi: Now that I think about it, maybe you're right that there's more to the past-future distinction than just some conventional selection of some hyperplane in somebody's favorite reference frame. Here's a thought. Let's reject a frame-relative present for a point-relative present that's more in line with the idea of proper time.

John: What does that mean?

Naomi: Well, by "point" I mean a spacetime point.

John: Does that mean there's a time for your point and a time for my point?

Naomi: Yeah, that's what proper time is, the time on the clock traveling with you along your world line. If you strip away the conventions, and look at the intrinsic geometry of Minkowski spacetime independent of any reference frame, then the only distinction between future and past required by the theory of relativity is the distinction between a single point's future light-cone and its past light-cone. The "hard core" past of a point event is all the events that in principle could have had some effect on the point. These points have the shape of a cone in the Minkowski diagram.

John: OK, I know a little bit about what a light-cone is, but every point has a different light-cone, right?

Naomi: Right. Neighboring points are going to have different light-cones, but their difference will be insignificant. Every living human being will have dinosaurs in their past cones. The disagreement comes in if I'm right now flying over your head very fast. You and I might disagree by a few seconds on when some explosion across the galaxy occurred, though we'd both be correct. But here's the main idea about dividing the future from the past. For a single point, even if it's possible to define the hyperplane of its present events relative to your favorite coordinate system, this plane won't divide the hard core future from the hard core past for the rest of us point-people—and it was this divider plane that Putnam was presuming. A single point's past cone differing from its future cone cap is what really divides the past from the future.

John: OK, this past-cone theory has some nice features. It is consistent with relativity theory, and it saves the idea that we almost always agree about the reality of past events and when they occurred. But I'd like more. I'd like the future cones to be unreal, and I want the present to be the nexus of becoming.

Naomi: I can see how you can hold onto your idea of becoming. Say that a point-relative reality bubbles into existence at each point. The bubbling is your flow of time.

John: Let's see if I've got this right. You're trying to describe a growing-universe theory with its dynamic theory of becoming. From this perspective, as time goes on, each spacetime point in the block gets included in more and more past light-cones of the other points. That's what you called "bubbling," and it's what becoming real is all about. I guess you're suggesting that if I believe in dynamic becoming I have to give up on saying there's a coherent notion of what is universally real at a specific time—universal in the sense of across the whole universe for everyone. I have to be satisfied with the notion of being real for a spacetime point. And I have to allow the idea that what is real at one point can be different from what is real at another. OK, I guess that's what I need to accept if I'm going to accept relativity theory, but I'm finding it difficult to hold onto my old views about what a dynamic flow is. The flow is fragmented into a bunch of growing light-cones all over a Minkowski diagram.

Naomi: I see your problem, but I don't have any suggestion for you. Besides, I never believed in dynamic flow in the first place.

John: I'll have to think about this problem. But let's change the subject a little. We still haven't talked about the other big issue in class today: whether the basic objects perdure.

Naomi: Perdure?

John: It's a technical term in metaphysics. The endure-perdure issue is about how objects exist over time. If an object perdures instead of endures, it can't exist wholly at an instant but requires a stretch of time to exist. Think of the event of singing a specific song in the shower from beginning to end. The whole song doesn't exist at any single instant. We normally think of a physical object such as our bus as enduring through time because we would say the whole bus exists now and also existed five minutes ago. The bus doesn't perdure. Enduring objects are three-dimensional; perduring objects are four-dimensional, assuming time is a fourth dimension. The first half of the song was a temporal stage of the song. If you think of spacetime as a four-dimensional block, then a temporal stage of a perduring object is a three-dimensional cross-section of the block from the stage's beginning time to its ending time. Perduring objects have temporal stages. Enduring objects don't. Instead, we say enduring objects "continue to exist," and metaphysicians call them "continuants." Classical ontology said the basic objects are 3-D enduring objects. Ontology influenced by Einstein says 4-D events are the basic objects, and they perdure rather than endure.

Naomi: Like in a Minkowski diagram; it's basically a diagram of events, not of 3-D objects. When I do physics I think of objects through time as just extended processes, what you might call 4-D objects.

John: OK, but think about whether *persons* perdure or endure. Did all of you exist in your childhood, or did only the childhood part of you exist then?

Naomi: Uh . . . that's a difficult question.

John: The background problem here is find out what a person is, and this requires knowing the criteria of identity for persons. We'd like clear identity criteria. They tell us about what makes a person be the same person over time, and which changes the person can or can't survive and still be the same person.

Naomi: Have you ever noticed how Superman and Clark Kent are never in the same room at one time?

John: Strange, isn't it?

Naomi: OK, I understand. It seems obvious to me that Americans perdure, but Canadians endure.

John: I was just about to say that.

Naomi: Seriously, what do you think about people perduring?

John: I believe I have spatial parts such as my elbow, but I'm reluctant to believe I have temporal parts such as my childhood. It's strange to call them parts. They're not part of my body, but I guess you can say they're part of my life. Which is the real me, my life or my body?

Naomi: I'd go with you being your life. That seems the right way to look at it. I think of a life as a process, a long event. Long events are the sum of short events. A temporal part of an event is just a shorter event, and parts of it are even shorter events, all the way down to an instantaneous part. The way I look at it, you and I are the sum of all our instantaneous temporal parts.

John: That's why philosophers call the perdurance theory "the doctrine of temporal parts." The doctrine is a very static view of people. It doesn't allow people to change, so that's a sign of trouble.

Naomi: Change doesn't so much disappear as get talked about in a different way, it seems to me. When an ordinary person says an enduring object changes, I'd just say that this should be analyzed as a difference in two temporal stages of the corresponding perduring object because one stage has a quality that the other doesn't have. Here's what I'm thinking. I bend my arm. The arm has a straight stage followed by a bent stage. I'm not denying change, just redescribing it. Change in x is nothing but difference in temporal stages of x. That's how I'd explain change with what you've called the doctrine of temporal parts, but I'm not sure how to argue that the doctrine is correct. It's easy to have an opinion; it's harder to argue for it. Did your class discuss arguments?

John: Yes. My friend Judith complained that the perduring-parts theory has a problem because the parts pop into existence as if by a miracle. She doesn't believe in miracles. The theory seemed to her to be a crazy metaphysic.

Naomi: It seems obvious to me that there's no miracle here. The previous temporal parts are causing the present temporal part, that's all.

John: That's what Mehlberg said. He said there's no inexplicable emergence going on. But someone else complained that the perdurance theory has a problem accounting for our moral responsibility. The theory needs to permit us to hold people and governments responsible for their actions. A murderer is responsible today for what he did last year, but why hold today's temporal part responsible for what last year's temporal part did since the two parts exist only for a short time and aren't identical?

Naomi: Can't you just say the whole person or whole government is guilty of murder if any of its temporal stages committed murder? Wholes are to blame for what their parts do.

John: Maybe, but it's a little odd to say a stage committed a murder. I'm thinking I walk into a bar, look the bartender in the eye, and say, "Bartender, give my temporal part a beer."

Naomi: Yeah, you'd get a weird stare and no beer. But you metaphysicians shouldn't worry too much about your theory being odd compared to ordinary talk. We physicists don't. Some particles inside a proton have charm and zero width. Some are up but not anti-down.

John: Up?

Naomi: Yeah, it's not a direction. Maybe we should have a contest to see which field says the oddest things, physics or metaphysics.

John: Could be fun. I'll describe the argument the student in class today gave in defense of the perdurance theory. Then you can tell me if it's a convincing argument. His idea was that the theory solves puzzles the theory of enduring objects can't solve, or can't solve as well.

Naomi: For example?

John: Back in Greece in the sixth century BCE, Heraclitus said that because you're stepping into different water, you can't step into the same river twice. After all, the water is essential to what makes it be a river. Leibniz's Law says if two things have different properties, essential or not, then they're different objects, so his law implies the river doesn't survive from one moment to the next. But we need the river to survive. That's the puzzle.

Naomi: And perdurance comes to the rescue?

John: Yes. Quit thinking of the river as a three-dimensional enduring object, the way we normally do, and consider it to be the four-dimensional sum of all its temporal stages. You step into different temporal stages but into the same sum of the stages, and it's the sum that is the river. That way you can have your change and your identity, too. You can live with Leibniz's Law forcing the temporal stages to be different objects. You don't need the temporal stages to be the same; you just need the sum to be the same sum. So there you have it: identity with change. You can't solve the Heraclitus puzzle by using enduring objects because that's how we got into the problem in the first place, at least according to the student who made the presentation today.

Naomi: OK, one point scored for perdurance.

John: Here's another metaphysical puzzle about time that needs a solution. Not only do we want a theory that allows objects to change while keeping their identity during the change, but also we need a theory that tells us what it is that makes the object before and after the change be the same object. On the theory of enduring objects, what makes a physical object the same object from one time to the next is usually said to be spatio-temporal continuity. Or if there's a little change from time to time, then identity is about continuity of replacement of substance and continuity of changes in the enduring object's shape or behavior or whatever. A cow walks from pasture to barn. It's the same cow because there's a continuous series of appropriately occupied space and time locations from pasture-then to barn-now even if the cow comes back muddier and hungrier. Now here's the puzzle. Suppose I take a lump of clay and mold it into a statue of a woman's body, then change my mind and mold it into a man's face. The woman statue is spatio-temporally continuous with the man statue, but they definitely aren't the same statue, are they? If you think of them all as enduring objects, then is the clay identical both to the woman's body and also to the man's face?

Naomi: No, that would make one equal two, and the mathematicians would complain. Hmm, this is really a puzzle.

John: The perdurance theory has an answer. Treat the clay, the woman's body, and the face all as perduring objects. The clay, the woman's body, and the man's face don't have the same stages, so none of them are identical to each other. But we still need to explain why the two statues are similar even if they're not identical.

Naomi: Oh, now I see. If they're perduring objects, then we can say the woman's body and the man's face are similar because both have stages that are stages of the clay. Yeah, that works.

John: One more point for perdurance. I still don't support the perdurance theory, but in class we learned about one more argument for perdurance: it gives the best solution to the problem of temporary properties. My arm is temporarily bent now, but when I stand up suppose I straighten it. I normally would think of my arm as an enduring object that is temporarily bent, then temporarily straight. For Leibniz, it can't be the same arm because the arm has different properties at the two times. But of course it's the same arm, so what do we do—throw out Leibniz's Law, or keep it and deny the possibility of change?

Naomi: Desperate moves, either way. No, wait a minute. Let's just revise Leibniz's Law and say that two objects are identical if they have the same *essential* properties. Your arm's being bent isn't essential to its being your arm. What is essential is that it's a solid and not a gas and that it's attached to the rest of your body and not attached to the top of Mount Everest.

John: OK, that's a way to treat change, but can you make sense of the distinction between essential and nonessential properties of enduring objects?

Naomi: Would this work? My arm being *necessarily* solid means that the sentence, "My arm is solid," is necessarily true. How's that? The sentence, "My arm is in Los Angeles," is contingently true and not necessarily true.

John: Well, how do you know that the sentence, "My arm is solid," is necessarily true?

Naomi: Because it's part of the definition of "arm." It's true by definition that an arm is solid.

John: OK, that's progress, but what justifies the definition? How do you know that solidity, say, is required by the definition but being in L.A. isn't?

Naomi: Because it follows from the meaning of the term "arm." I know you're immediately going to ask me to justify this justification. If you do, I'm not sure what to say other than that's how we use the words. If you ask me why we use words this way, I'd like to say, "We talk this way because being solid is essential to my arm," but I know I'd be arguing in a circle if I said that.

John: At least it would be a *big* circle. You've connected being essential with necessary, and necessarily true, and true by definition, and true by the meaning of the terms, and back to essential again. Quine argued in the 1950s that it's hopeless to escape the circle. We were hoping to solve the problem of temporary properties by letting the arm be an enduring object that changed its inessential properties temporarily. Then you wanted to revise Leibniz's Law to be about essential properties and say the two arms are identical not because they have all the same properties but rather all the same essential properties. That solution didn't work out. So let's look at another solution. According to the student who was arguing for perduring objects, the solution is to retain Leibniz's Law as is but give up on thinking of an arm as being an enduring object that gains and loses the qualities of being-bent and being-straight. It's a perduring object that is neither bent nor straight. Instead, say the arm's earlier temporal part has a bent-arm quality, and the arm's later temporal part has a straight-arm quality. I know I'm inventing some new qualities for perduring objects. Now the temporal parts are different according to Leibniz's Law, but who needs the two temporal parts to be the same anyway? If my arm is a perduring object, then the sum of its temporal stages doesn't change when I perform the act that we informally call straightening my arm.

Naomi: Victory again for perdurance.

John: Well, there you are. You have all these puzzles with enduring objects. There's Heraclitus's problem, the problem with the lump of clay, and the problem of temporary properties, and you can solve them best with the thesis of perduring objects, and you don't have to tinker with Leibniz's Law, and you don't have to deny the possibility of identity with change; so you should conclude that objects perdure rather than endure. If you accept a four-dimensional world of temporal stages, you don't get into metaphysical trouble. You asked for an argument. That was the student's argument.

Naomi: I'm now convinced everything perdures.

John: *I'm* not.

Naomi: What?

John: That's right. Perdurance doesn't really solve the problem of what it is that gives something its identity through change; the problem is still there in disguise.

Naomi: It is?

John: First of all, even when we treat the arm as a perduring object we have to decide what unites all the temporal stages so that they're stages of the arm rather than of a balloon. Saying the stages have the "right relationships" or are "appropriately related" just hides the problem. What makes a relationship the right one? If one of the temporal stages is a gas instead of a solid, then we'll all agree the arm suddenly stopped existing. Why? Because it wasn't "right"? No, because being solid is *essential* to being an arm. So the old problem is still with us. When you change something's inessential properties, you want it to survive the change. When you change something's essential properties, you don't want it to survive. Do you see the problem?

Naomi: Yes.

John: Well, that's one reason I'm not convinced that the perduring theory is the way to go. Second, I don't like thinking of all objects as really being processes. The implication is that some sentence I thought was about an enduring object such as "Lincoln is president" has to be analyzed in terms of processes as "Lincolnizing presidentially is going on." The processes need new qualities so I'm using the new quality of "presidentially" for the perduring object. Doesn't this have all the smell of some mistake made in a metaphysical laboratory?

Naomi: It smells a little odd to me, too.

John: A world of pure processes is so strange, so unintuitive, that it makes me suspicious. And, besides, I can think of another way to solve all those puzzles.

Naomi: You can?

John: Yes, stick with enduring objects but use a more sophisticated concept of staying the same. Heraclitus's river is not the same sort of river from one step to the next if the river is considered to be a pile of water molecules, but it *is* the same sort of river when it's considered to be a water-filled geographical landmark. Identity over time of an object should always be meant relative to some category or way of considering what the object is. See? Puzzle solved.

Naomi: OK, that's a clever way out. Identity can be relative to your purpose. Not declaring your purpose and simply asking, "Can you step into the same river twice?" is asking an ambiguous question.

John: Exactly. Thanks for helping me develop my idea.

Naomi: You're welcome. How are you going to handle the bent-straight problem?

John: In class today, I recommended a shift from qualities to relationships. Give up on using the qualities bent and straight, or the new qualities of bent-arm and straight-arm, and instead use a new relationship, the bent-relationship that holds of my enduring arm and a time. It holds at some times but not at other times. No need to bring in perdurance.

Naomi: So you're rejecting the simpler qualities of being bent or straight and analyzing the bent-straight problem using complicated new relationships.

John: I know. Wait. On second thought, I don't really want to give up on the simple quality of being straight. So how about this? Nobody in class mentioned it. Declare that my enduring arm is-at-time-t bent and is-at-time-t straight. The simple quality of being straight is back. It's the *having* of the quality that's time-relative. How's that? It's not common sense, but it does the least damage to common sense, I hope.

Naomi: That's reasonable. I'll have to think about the implications of this, but I don't see why I couldn't say "perduring" everywhere that you were saying "enduring." One thing I've learned from this conversation is that solving these problems about what change is, and what a person is, can't be solved simply by probing the ordinary speaker's meanings of those terms.

John: Yeah, instead, the meaning is up for grabs in the struggle to resolve the conflicts among metaphysical assumptions, intuitions, meanings, and scientific knowledge. A delicate balancing act. Well, my stop is coming up again.

Naomi: See you next . . . Oh, that's right; it's Thanksgiving vacation. OK, after that.

John: Are you going to be with your family?

Naomi: I'll be there the whole time, and I'm hoping to see some friends who'll be back from college.

John: Great. After two days with Mom, Dad, and Mr. Turkey, I'm off to a lake. One of my friends' parents owns this house on the shore, and seven or eight of my hometown friends are going to spend the weekend there. Should be an awesome party.

Naomi: Sounds like fun.

John: Hope so. Bye.

DISCUSSION QUESTIONS

1. How might a presentist account for our meaningful talk about the past if the past does not exist to be talked about?
2. What do you think about the remark that the present is the fountain where the river of time gushes out of nothingness? Why?
3. Summarize Naomi's reasons in favor of the block universe.
4. The law of bivalence says every statement is either true or false. Is there evidence that this law fails for any statements about time?
5. Summarize in your own words the Putnam argument and the response that Naomi recommends. If John accepts it, then how does his growing-universe theory have to be revised?
6. How about saying that a chair both perdures and endures and metaphysicians should quit trying to decide between the two? Couldn't the chair perdure for some purposes and endure for others?
7. Are you your worldline?
8. Could you have lived in another time, say the fourteenth century, and still be you? Why would others disagree with you?
9. Suppose your left arm is bent at one time and straight at another. Apparently by Leibniz's law it's not the same arm from one time to the next. What's the best way to resolve this problem?
10. Not only do we want a metaphysical theory that allows people to change while keeping their identity during the change, but also we need a theory that tells us what it is that makes them before and

after the change be the same person. What's the best solution to these two problems?

FURTHER READING

Broad, C. D. *Scientific Thought*. London: Kegan Paul, 1923.
The classic statement of the growing-universe ontology of time.
Chisholm, Roderick M. "Identity through Time." In *Language, Belief, and Metaphysics*, edited by H. E. Kiefer and M. K. Munitz, 163–82. Albany: SUNY Press, 1970.
Discusses the Ship of Theseus and the identity of persons who endure through time.
Dainton, Barry. *Time and Space*. Montreal: McGill-Queen's University Press, 2001.
For a good discussion of the block universe, with pictures, see "block universe" in the index.
Gallois, Edward. "Identity over Time." *Stanford Encyclopedia of Philosophy*.
Survey of the philosophical controversy about retaining identity through a change.
Hawking, Stephen. *A Brief History of Time: The Updated and Expanded Tenth Anniversary Edition*. New York: Bantam Books, 1998.
In chapter 2, this important physicist discusses what spacetime is and why he believes in it.
Le Poidevin, Robin. "Change." In *Routledge Encyclopedia of Philosophy*, edited by Edward Craig. London: Routledge, 1998.
Explores various views on how change is to be understood and analyzed.
———. "Continuants." In *Routledge Encyclopedia of Philosophy*, edited by Edward Craig. London: Routledge, 1998.
Introduces some arguments for and against enduring objects and perduring objects.
Lowe, E. J. "The Problems of Intrinsic Change: Rejoinder to Lewis." *Analysis* 48, no. 2 (1988): 72–77.
Lowe attacks David Lewis's idea that your arm perdures rather than endures. Lowe says your arm is-at-t bent and your arm is-at-t straight.
Noonan, Harold. "Identity." *The Stanford Encyclopedia of Philosophy*.
Surveys the issues about perdurance for many sorts of entities, including persons.
Parfit, Derek. *Reasons and Persons*. Oxford: Oxford University Press, 1984.
Parfit bites the bullet and says that when a person fissions into two, both are the original person, despite the violation this does to our common sense.
Prior, Arthur N. "The Notion of the Present." *Studium Generale* 23 (1970): 245–48.
A brief defense of presentism and an elementary introduction to tense logic. Prior reacts to the Putnam (1967) claim that presentism is inconsistent with special relativity by claiming that this reveals an imperfection in special relativity.

Putnam, Hilary. "Time and Physical Geometry." *Journal of Philosophy* 64 (1967): 240–47.

Argues that Einstein's special theory of relativity is inconsistent with the unreality of the future and by implication with the idea that dynamic becoming and the flow of time are a matter of future events becoming real.

Quine, W. V. O. "Two Dogmas of Empiricism." In *From a Logical Point of View: 9 Logico-Philosophical Essays*, by W. V. O. Quine. Cambridge, MA: Harvard University Press, 1953.

Quine argues that there's no noncircular way to distinguish statements true by definition from those that are contingently true.

Sellars, Wilfrid. "Naturalism and Process." *The Monist* 64 (1981): 37–65.

Examines the suggestion that the world consists solely of processes rather than enduring objects. In terms of semantics, this means treating "Lincoln is president" as "Lincolnizing presidentially is going on."

Sider, Theodore. *Four-Dimensionalism: An Ontology of Persistence and Time*. Oxford: Clarendon Press, 2001.

A defense of the perdurance theory.

Smith, Quentin, and Nathan Oaklander. *Time, Change and Freedom*. London: Routledge, 1995.

Explores how belief in the reality of the future affects belief in free will.

Stein, Howard. "On Einstein-Minkowski Space-Time." *Journal of Philosophy* 65 (1968): 5–23.

An influential response to Putnam (1967). Stein says, "In Einstein-Minkowski space-time an event's present is constituted by itself alone." The response is expressed more simply in Dainton (2001), 272–77 and 283, and Sider (2001), 42–52.

Zimmerman, Dean W. "Temporary Intrinsics and Presentism." In *Metaphysics: The Big Questions*, by Peter van Inwagen and Dean W. Zimmerman, 206–19. Malden: Blackwell Publishers, 1998.

Criticizes David Lewis's 1986 argument that embracing temporal parts is the best way to solve the problem of temporary properties.

8

The Arrow of Time

John: Welcome back from vacation.

Naomi: Thanks. Are you still eating leftover turkey sandwiches like me?

John: No, I'm on an all-chocolate diet. It's not as effective as exercise, but I figured it wouldn't hurt to try.

Naomi: Good luck with that. My mom sent me back to L.A. with a ton of leftover turkey like she does every year. I had so much in my suitcase I probably made the airport's bomb-sniffing dogs go crazy. But anyway, how about you? Did you enjoy your vacation?

John: Yeah, but it's wasn't as good as when I was nine years old.

Naomi: How's that?

John: When I was nine, I caught my first fish using a rod that was only two feet long. My very first cast was a huge success. I hauled in a thirteen-inch trout.

Naomi: What luck!

John: When I was in the boat on the river last weekend, I was dreaming of a repeat. Didn't happen. But I had fun anyway. While I was there on the bank watching things float downstream, it reminded me of being in time with the events of my life passing by.

Naomi: Those events never stop coming. The arrow of time points up-stream, but did you ever think about why it points that way instead of the other way around? Why do you think the arrow points toward the future instead of the past?

John: It's just a conventional truth. What's not conventional is that the universe behaves as if it has an arrow.

Naomi: Without an arrow, there'd be chaos.

John: Definitely. In class today one of the students said the arrow should point the way it does independently of how events go in the world. Other students complained that he didn't understand the arrow.

Naomi: I'm not surprised. Most people would say the arrow depends on events happening—first-order changes, as you call them.

John: Mehlberg said one of the big problems with the direction of time is to figure out just what the problem is. Some people say the problem is to show that there's an intrinsic difference between time flowing to the future and time flowing backward to the past, but other people say that's not the problem because we simply define "future" as how time flows, and the real problem is to decide why there's asymmetry. Others object to describing the arrow with the word "flow" because they believe in the arrow but not in its describing the direction of flow.

Naomi: I'm one of those people who don't like the word "flow," but what did you mean just now by "intrinsic difference" in the two directions?

John: Oh, like a leaf's being green is an intrinsic quality of the leaf. Its being west of New York isn't, because that's just a relationship the leaf has with some other thing. There's an intrinsic difference between its being green and being brown. And the question is whether there's an intrinsic difference between time flow to the future and time flow to the past.

Naomi: OK, that's what I thought "intrinsic difference" meant.

John: So what do physicists say about the arrow?

Naomi: That there are arrows, and then there are arrows. The arrow of a specific physical process is the way it normally goes. A lit wooden match normally goes from wooden match to ashes and never from ashes to unburned wooden match. If a process can't be reversed without altering the surroundings, physicists call it an "irreversible process." Imagine what you'd have to do to the surroundings to produce a wooden match starting with ashes. Most processes you can think of are irreversible. The amalgamation of the universe's irreversible processes produces the cosmic arrow of time, the master arrow. Usually this arrow is what we mean when we say "time's arrow." And physicists would say it exists regardless of whether you accept the substantivalist theory of time or the relational theory.

John: Yeah, but even reversible processes are changes in time, aren't they?

Naomi: True, but they can't be used to define an arrow. The process of several molecules bouncing off each other is a good example of a reversible process. Here's why. Suppose during a microsecond I were able to take a series of photographs of the molecules as they collide and bounce away. Then I shuffle the photographs and ask you to unshuffle them. You could order them properly, but there'd be two sequences to choose from, the sequence with time going forward and the one with time going backward. Which to choose? That indecision is a sign that the arrow of time can't be found in this process. You have to look for the arrow on a bigger scale.

John: OK, the arrow isn't detectable by examining reversible processes for short time periods, but surely the arrow is still there even if it can't be detected. You believe things can exist without them being detected, don't you?

Naomi: Yes, but if it's never, ever detectable, why believe it's there? If somebody tells you they have an invisible friend who's a rabbit, but who can't ever be detected by you, then why believe them?

John: Hmm . . . OK, so your point is that it's not that time fades away in small regions for short times, but just that the arrow does. And you're saying the arrow of processes is a collective property that emerges only on a larger scale in processes involving many particles.

Naomi: That's a good way to put it. You'd think that an inspection of the laws of nature should reveal why directed processes go one direction and not the other. Well, when it comes to the very basic physical laws, nobody can find an arrow.

John: When a match burns, doesn't the law of match burning say that over time there has to be more charcoal rather than less, so isn't that a direction inside the law?

Naomi: Yes, but you're looking at the situation too macroscopically. At the microscopic level where we have more basic laws about molecules, the laws explaining burning allow for the process to go either way. The match can unburn. Essentially, all the basic laws of fundamental processes are time symmetric. This means that if a certain process is allowed by the laws, then that process reversed in time is also allowed, and either direction is as probable as the other, as far as you could tell just from the laws themselves. Since the history of the world is the history of all those processes, this means that if the world has one history, then it could have had the reverse history.

John: That's odd. If the laws of science don't make time have the arrow it does, then I wonder if this shows the laws are defective.

Naomi: Maybe it does. Or maybe it shows that time doesn't have an intrinsic arrow after all.

John: But how about entropy? I don't really have a good sense of what it is except that it's some measure of the disorganization of a system, but isn't it associated with the arrow? I mean, doesn't the second law of thermodynamics require an increase in entropy over time for isolated systems, so isn't entropy's overall increase the same thing as time's arrow? In your class last year we learned that increase in entropy is why matches burn but never unburn, why cans rust but never unrust, and why engines are never perfectly efficient. Is that right?

Naomi: Yes. This law does describe the arrow, but it's not fundamental. It's really a law that can be explained in terms of the statistics of molecular processes that are more fundamental. If you have an isolated system of just a few particles, then the processes involving those particles can go either way, but if you have zillions of particles, then the more probable direction reveals itself. It's not that it can't go the other way but that it won't. The high probability of a process going one direction rather than the other is what makes entropy very probably increase, and it's what makes us say the process is "irreversible." But the fundamental laws are always reversible; that means, if the process went in the improbable direction, then the process could still be obeying the same laws.

John: I think I need an example.

Naomi: OK, instead of thinking about a chemical process, let's think about processes involving the mixing of air molecules of different speeds. Suppose we have an isolated box with hot air in its left half and cold air in its right half. The box is very ordered, we say. As the molecular particles interact by bouncing off each other, there are many possibilities for a final distribution of the molecules ten minutes later, but most of the possibilities are ones in which the molecules are nearly evenly distributed. in both halves and the original order is gone. The most probable distribution after ten minutes is one in which you have molecules of the same speed distributed uniformly throughout the box. It would be extremely improbable to find after ten minutes that the left half of the box was even hotter than when we started. The measure of this directedness toward the probable is what physicists call "entropy's increase." It's the direction of increased disorder. That's the thermodynamic arrow in action. To detect the arrow, you can't look at only a hundred molecules. You have to move out to the level of generality of millions of molecules. These molecular mixing processes go the way they do because nature is producing probable states from improbable states.

John: So that's what physicists think the arrow of time is? Some urge toward the probable? Entropy change is a measure of this urge?

Naomi: Yes. Well, almost. That's just the thermodynamic arrow because entropy change is an idea that applies only to situations involving energy flow that can be measured as heat change. It's a big part of the story of the cosmic arrow, but there are other arrows of time that also point to the future and seem to have nothing to do with heat change. Effects happening after their causes and not before is called the "causal arrow." In order for this causal arrow to exist, normal causation has to be much more frequent than backward causation. The psychological arrow is the direction from our sensing of events to our having memories of those events. The psychological arrow is what marks our remembering the past rather than the future. Another arrow exists when possibilities decrease and actualities increase as time goes on, but I don't have a name for that arrow. Clusters of galaxies are receding from one another on average as time goes on; this is the cosmological arrow that indicates the universe's expansion. Light going out from a candle flame rather than converging into the flame indicates the electromagnetic arrow. And there are many other arrows. For most of the arrows, the physicists don't know why the arrows exist, but just that they do. Nobody knows why light doesn't converge into the flame instead of only shining out from it.

John: I'm really surprised that scientists don't know why.

Naomi: Science has a ways to go. Physicists don't know how the universe acquired its initial low entropy, and they don't know whether one arrow is more fundamental than the others or if they're linked somehow. I worry that maybe they aren't linked. Then wouldn't it be strange if no other arrow changed except that the psychological arrow reversed tomorrow morning and we started having true memories of events before we experienced them?

John: It'd be really strange.

Naomi: I know you'd disagree with me, but I'd say the arrow of time is objective and the feeling of time's flow is subjective. They both point the same direction because the feeling of flow is due to changing brain states, such as acquiring new memories, and these changing brain states are thermodynamic processes that move toward higher entropy just like all the other processes. This is why the direction that time seems to flow is also the direction the thermodynamic arrow actually points. So the thermodynamic arrow is more fundamental than the psychological arrow. How about that?

John: That would make sense if the flow were subjective and due to brain processes, but I still think it's objective. Maybe it's time's flow that forces time's arrow to have the direction it does. Anyway, think about all the processes reversing. I don't know how, but just think of it. Then the cosmic arrow would reverse and time would run backward. We'd be able to relive our past. Like, look, up ahead, there's the past! It would be like when you step out of a time machine and see the past.

Naomi: I don't think the experience of arrow reversal would be like the experience of time travel. With time travel, the past is revisited in the original order that the events occurred. With arrow reversal, the past is visited in reverse order, which would be much stranger, don't you think?

John: Yes, stranger. I guess you're right about time travel not being like time reversal. OK, but suppose the universe were divided into two regions, one like ours with a normal arrow and one with the arrow reversed from the way it goes in our region. If people there were built just like us, with the same kinds of bodies and minds and civilization, then they'd usually walk backward up stairs, and food would come out of their mouths when they ate. Their past would look like our future.

Naomi: Wouldn't you say that their future, like ours, is by definition what will happen, not what has happened?

John: Yes. If we were able to watch them in their region, they might appear to us to be precognitive. I can imagine how they could easily win gambling bets on the roll of the dice. The un-roll.

Naomi: I'd have agreed with you about gambling a minute ago, but now I don't think someone's experience of a time-reversed world would be strange at all compared to our experience. Even if the people in the time-reversed region did know the numbers on the dice before they were rolled, their experience of dice wouldn't be different from ours because their brain processes would be reversed, too.

John: Uh . . . The reversal of their brain processes wouldn't remove the fact that they'd have access to the result of the roll before they placed their bet on the roll. That's all they'd need to easily win the bet. Reversing the cosmic arrow would require people to remember events in their future, and the future is what hasn't happened yet.

Naomi: Who'd accept the bet if it was common knowledge what the outcome is? Maybe there couldn't be true gambling as we know it.

John: It's so hard to imagine what their experience would be like. Mehlberg said that if we lived in a time-reversed world, we'd discover the same scientific laws we have now. Really odd.

Naomi: Different arrow; same science.

John: If Aristotle is correct that our future is undetermined or open, then because our future is like their past, the past of people in the time-reversed region would be open, too. And if their past could be open, why not ours?

Naomi: That's a dangerous deduction.

John: I know. Do you think we could communicate with the time-reversed people? If we sent a signal, could our message cross the border, or would it dissolve there, or would it bounce back? I'm not sure what would happen. I guess all the molecules involved in the signal would reverse direction at the border, and that should be enough to make their region impenetrable, like a perfect mirror. Odd.

Naomi: Just because the processes in the other region are the reverse of ours, that wouldn't make our own processes go in reverse as they arrived at the border. I think our signals would simply enter an odd, dark world where light is converging into lamps and people in the cemetery are rising out of their coffins and becoming undead.

John: Yeah, I think I was wrong about that mirror idea. If we were able to send a message to an inhabitant of the time-reversed region, maybe they couldn't remember our message because it would become part of their future. No, they remember the future, don't they?

Naomi: Memory is always about retaining traces of what's happened. So the message would enter their past while they played it. They'd just need to play the message backward, that's all.

John: No, that definition of memory you're using works only in our own region.

Naomi: I'm not sure.

John: How about this? Forget about regions. Suppose the whole universe contained people who were just about as good with foreseeing experiences as with remembering experiences. They have noninferential, direct knowledge of their future experiences just as we do of our past experiences. Their world wouldn't have any psychological arrow of time, would it?

Naomi: First off, you should distinguish whether they could or couldn't tell the difference between memory and precognition. Suppose they couldn't. Their world would still have the usual cosmic arrow, but they wouldn't be any good at detecting it naturally. We normal people use our memory and our ignorance of the future to detect the arrow and to distinguish between before and after. You know, if, while you're eating the apple, you truly remember buying the apple, then the buying was first and the eating is second. If we lost this detector of the arrow, that wouldn't wipe out the arrow itself.

John: OK, it wouldn't wipe out the cosmic arrow, but what about the psychological arrow?

Naomi: I'm not sure, but here's a thought. Abnormal people who are as good with the future as the past, but can't immediately tell the difference, could build an entropy detector and then rely on entropy increases to tell them which direction the future is. More generally, when these abnormal people are faced with two candidates for time order for a sequence of events and both orders look natural to them, they would find that only one of the two was consistent with the second law of thermodynamics.

John: Imagine living your life relying on an entropy detector machine to work out whether your mental image of falling in a mud puddle is a thought of what might happen or instead of what has already happened.

Naomi: An insane way to live.

John: Do you think the arrow in our normal universe can reverse? I don't see why not.

Naomi: It can't happen. I mean it can't happen everywhere. The cosmic arrow can disappear but not reverse. If causality goes wild, then it will disappear, but it isn't allowed to reverse.

John: Why not?

Naomi: Having the whole universe reverse all its processes at once would be like having the whole universe change into its mirror image. There'd really be no difference. If all there is in the universe is a single hand, then it can't be either a right hand or a left hand. The difference between left and right wouldn't make sense. So, if all the processes reversed, it wouldn't make any sense to say the universe had a right hand but then everything reversed and it turned into a left hand.

John: You just made sense saying it, so why did you say it doesn't make sense?

Naomi: I was speaking loosely. I should have said there'd be no detectable difference, so by Leibniz's Law there'd really be no difference.

John: God would know the difference, Leibniz or no Leibniz. Also, I can think of how the cosmic arrow could reverse and there'd be a detectable difference. Imagine that our universe contains nothing but people who shuffle playing cards. Their arrow of time shows itself in the fact that when people begin with a deck in perfect order, a few shuffles will always destroy that order. Continued shuffling won't restore the order. But now suppose one day the shufflers were to start noticing that in a few shuffles they could go from a disordered deck to a perfectly ordered one. Then they'd be justified in saying, "Look, the arrow of time must have reversed today!"

Naomi: I can understand what you're saying, but this scenario only works because you're imagining the people in the universe not being involved in those processes that reverse.

John: I don't see the problem. OK, let the people reverse, too. Let *all* the processes in the universe stop and then start going the other way.

Naomi: What do you mean by "then"?

John: Well, I . . . OK, you got me there. . . . No, you don't! Time can go forward even though the arrow reverses.

Naomi: Now you've got two arrows. You're imagining God's absolute arrow of time going forward while he reverses the arrow of time of his creation.

John: That's right.

Naomi: But without absolute time, the reversal would be described very differently by someone living during the reversal. I'd guess they'd say the arrow disappeared but is back again. They'd never notice the reversal because it would be analogous to our not noticing our waking up into a mirror-reversed world.

John: Are you distinguishing between saying people couldn't notice the reversal and saying there's no reversal?

Naomi: I made a mistake and forgot to make that distinction, but now that I think about it, if a reversal is not detectable even in principle, then it makes no sense to say there's been a reversal. It's all about Leibniz's Law of the identity of indiscernibles. A mirror-reversed world remains the same world.

John: I'm not sure. We seem just to be clarifying our difference rather than removing our difference.

Naomi: Well, we still have the issue of whether there's any physical property of the universe that gives time its direction. I think the property is causation. I want to define directed time order as causal order, but now I'm worried that we can't explain what causal order is without presupposing we already know what time order is. Maybe I'm going in circles. Time is really difficult to think about. Space is so much easier.

John: Mehlberg mentioned that Hans Reichenbach had an original idea about this back in 1924. Two events can be put in the proper time order if you can figure out which one of them could have caused the other, but not vice versa. Reichenbach used that idea, and his definition was something like this. Event A is earlier than event B if either A was part of the cause of B, or an event simultaneous with A was part of the cause of B, but not vice versa.

Naomi: That's fine, but I think Reichenbach's definition doesn't work if there's a possibility of backward causation.

John: He probably ruled that out.

Naomi: I'm still waiting for a good reason for doing this.

John: OK, but putting that aside where are we? The idea is that A happens before B if A could have caused B. Now Reichenbach's next problem is to avoid circularity by somehow defining what it means for A to cause B without presupposing that, if A could have caused B, then A must happen before B.

Naomi: This won't be easy. We know the laws of science are time symmetrical, which means that in principle if A is followed by B, then B can be followed by A. If a hammer dropping on an egg can be followed by the egg cracking, then in principle the egg cracking can be followed by the egg being whole and the hammer rising. I suppose somewhere Reichenbach has to start talking about statistics or else give up on an objective description and say causation is something the mind projects onto reality.

John: His tactic here does involve statistics. He defines causal order in terms of an asymmetry due to outgoing processes from a common center tending to be correlated with one another, but incoming processes to a common center tending not to be correlated with one another.

Naomi: Outgoing and incoming?

John: Yeah. In class we talked about an example used by David Papineau. We know that obesity and high excitement are two causes of heart attacks. These are two incoming processes tending toward a common center of heart attacks. But these two causes are probabilistically uncorrelated in the sense that obese people aren't any more likely to be highly excited than nonobese people and highly excited people aren't any more likely to be obese than unexcited people. Now look at outgoing effects from smoking. Both lung cancer and having nicotine-stained fingers are effects of smoking. This causation *does* imply that lung cancer is correlated with nicotine-stained fingers; for example, lung cancer is more likely among people with nicotine-stained fingers than among people with ordinary fingers. So the two causes of heart attacks are uncorrelated, but the two effects of smoking are correlated. This is a clue to the difference between what causes are like *versus* what effects are like. Reichenbach's idea is that effects can be detected by their being correlated or probabilistically dependent on each other, unlike causes. Reichenbach's program is considerably more complicated than what I've just said, but that's his key idea for distinguishing causes from effects.

Naomi: Reichenbach is onto something.

John: I'm not so sure of that. It seems to me that there are better ways to distinguish causes from effects than in terms of probabilities and correlations. I think it's best to think of causes as the sorts of events you manipulate in order to get the effects. You can't manipulate the effects to bring about their causes. That's what really distinguishes causes from effects, not probabilities and correlations. Think about kinds of events instead of single events. One kind of event is the firings of a gun at targets on the practice range. The other kind of event is holes appearing in targets. How do you tell which causes which without making the decision by saying the cause is what happens first? Well, you can't manipulate the target to get the gun to fire, but you sure can manipulate the gun to get holes in those targets. That tells you how to distinguish causes from effects, and then you can declare by definition that the causes are first, that is, before their effects. So there's your direction of time, the overall flow from causes to effects. I know this is vague, but that's my general idea.

Naomi: Did you escape circularity? I think you might be surreptitiously presupposing you already know which occurred first. The holes in the targets might be stimulating the pleasure centers of gun owners so they go and point their guns at the targets. The holes might cause the firings.

John: You couldn't say that about a single event. A single bullet hole in the target couldn't stimulate someone to shoot the bullet.

Naomi: Why not?

John: Because we know that first you pull the trigger, and then the bullet . . .

Naomi: Oops! You said "first" and "then." You've just assumed you know the direction of time, so there's a circularity in your theory.

John: All right, but maybe I can revise my theory. How about this? Suppose some event has happened. You can explain it by citing its cause; you can't explain it by citing its effect, and this asymmetry in explanation might be the key to distinguishing causes from effects—to their being different, not just to our noticing that they're different. Back to the gun and the targets. You want to explain the holes in the targets. Well, the gun firings are what explain the holes, so they're the cause.

Naomi: No, you've got the same problem. How can you know not to explain the firing of the gun by the appearance of the bullet hole?

John: Well, the explanation would be so bizarre and unintuitive. . . . Oh, there's my bus stop.

Naomi: Time flies.

John: As they say, time flies like an arrow, and fruit flies like a banana.

Naomi: [laughs]

John: See you next week.

DISCUSSION QUESTIONS

1. How do we know time's arrow doesn't point toward the past?
2. What's the difference between the arrows of time and the arrow of time?
3. Why do we remember the past but not the future?
4. What does entropy have to do with the arrow of time?
5. If the cosmic arrow of time went one way in our half the universe and the other way in the other half, and you received a verbal communication from the other region (presumably in English), what would your experience of it be like?
6. Discuss seriously the comments in the following dialogue between Alice in Wonderland and the Queen who lives backward:

   "Living backward!" Alice repeated in great astonishment. "I never heard of such a thing?"

   "–But there's one great advantage in it, that one's memory works both ways," said the Queen.

   "I'm sure *mine* only works one way," Alice remarked. "I can't remember things before they happen."

   "It's a poor sort of memory that only works backward," the Queen remarked.

   "What sort of things do *you* remember best?" Alice ventured to ask.

   "Oh, things that happened the week after next," the Queen replied in a careless tone.

7. How can causes be distinguished from effects?
8. What point is Reichenbach making in his discussion of the difference between incoming and outgoing processes and their correlations?
9. Is anything preventing the arrow of time from reversing?

FURTHER READING

Davies, Paul. *About Time: Einstein's Unfinished Revolution.* New York: Simon & Schuster, 1995.
A physicist surveys the problem of time's arrow in chapter 9.

Gardner, Martin. "Can Time Go Backward?" *Scientific American* 216, no. 1 (1967): 98–108.
An informal exploration of how the reversal of time's arrow is treated in the literature of physics, science fiction, and philosophy.

Hawking, Stephen. *A Brief History of Time: The Updated and Expanded Tenth Anniversary Edition.* New York: Bantam Books, 1998.
Chapter 9, "The Arrow of Time," discusses why all the arrows of time point in the same direction. Hawking explores what experience is like in a time-reversed world. He claims "the reason we observe this thermodynamic arrow to agree with the cosmological arrow is that intelligent beings can exist only in the expanding phase."

Horwich, Paul. *Asymmetries in Time: Problems in the Philosophy of Science.* Cambridge, MA: The MIT Press, 1987.
A philosophical monograph devoted to the problem of understanding asymmetries in time and to the relationship of this problem to other problems in philosophy. Our dialogue lightly touches on the relationship between thermodynamics and entropy, but it is covered well and accessibly in Horwich's chapter 4.

Morris, Richard. *The Big Questions: Probing the Promise and Limits of Science.* New York: Henry Holt and Company, LLC, 2002.
Pages 12–28 contain a brief, easy-to-read explanation of current scientific knowledge about the arrows of time.

Newton-Smith, W. H. *The Structure of Time.* London: Routledge & Kegan Paul Books Ltd., 1980.
See chapter 9 for this philosopher's introduction to the problems involving the direction of time.

Papineau, David. "Philosophy of Science." In *The Blackwell Companion to Philosophy*, edited by Nicholas Bunnin and E. P. Tsui-James, 290–324. Oxford: Blackwell Publishers Limited, 1996.
See page 312 for how two causes of heart attacks are uncorrelated, but two effects of smoking are correlated, which is a sign of how to distinguish causes from effects in general.

Price, Huw. *Time's Arrow and Archimedes' Point: New Directions for the Physics of Time.* New York: Oxford University Press, Inc., 1996.
An analysis of time's arrow from a distant vantage point outside of time. Difficult reading.

Reichenbach, Hans. *The Direction of Time*. Berkeley: University of California Press, 1956.

An original attempt to explain why there are traces of the past but not of the future. He develops his causal theory of time and speculates about there being two regions of the universe with their arrows of time going in opposite directions. Difficult reading.

Sklar, Lawrence. "Up and Down, Left and Right, Past and Future." *Nous* 15 (1981): 111–29.

Argues that the meaning of the term "arrow of time" is not what needs to be explained; rather there should be a theoretical reduction of the arrow. Similarly, earlier scientists didn't want to know what the term "water" means, but they wanted a theoretical reduction of water to H_2O.

9

Zeno's Paradoxes
and Supertasks

John: Hey, welcome back.

Naomi: How's it going? What are you listening to?

John: Beethoven's Ninth Symphony, his last.

Naomi: Yeah, it's famous. He was deaf by then.

John: The European Union rearranged a part of it for their official anthem, and now it's patriotic music.

Naomi: I didn't realize. Good choice for the anthem.

John: So how's your day been?

Naomi: Great!

John: Let me guess. The look on your face says you cut class to go to a monster-truck rally.

Naomi: A near miss. I finally caught up on all my reading for my courses.

John: Now you can relax for awhile.

Naomi: Yeah, but it's relaxing to be able to sit and read about interesting topics. Today was a special day for you, too. It was the day for your Zeno presentation. How'd it go?

John: Well, we had a metaphysical battle, but it started out calm enough. Back in Greece in the fifth century BCE, Zeno began the grand shift away from using poetry in philosophy and toward using prose that contains explicit reasons and conclusions.

Naomi: I never realized somebody deserved credit for that.

John: He's the man. He wrote a book of forty paradoxes, but only a few sentences survived. The last copy of the book is probably in a jar underneath an Italian freeway. Historians have been able to reconstruct nine of the paradoxes from comments made by his critics. I introduced the first three mentioned by Aristotle, then described the standard solution and his solution.

Naomi: Whose solution, Zeno's or Aristotle's?

John: Aristotle's. Zeno himself thought they show that motion is impossible, and hardly anybody followed him down that path. The three paradoxes are the Dichotomy Paradox, the Achilles Paradox, and the Arrow Paradox. They're paradoxes of motion. Because motion is a good example of change through time, he's really attacking our understanding of both change and time. The Achilles is the most well known. Zeno knows Achilles can catch a tortoise that crawls slowly away from him. On the other hand, Zeno argues that Achilles can't catch it. Here's why. His first goal is to run to where the tortoise is. When he gets there, the tortoise will have crawled off to a new place, so Achilles will have a new goal, to run to this new place. But he'll find the tortoise has crawled on again, and so forth. Although the gap between Achilles and the tortoise steadily decreases, there's no final goal, so he'll never catch the tortoise. So much the worse for our belief that motion really occurs. So what are we to do with this contradiction? Nobody thinks we can escape the paradox by jumping up from our seat and chasing down a tortoise. Even Zeno agrees that we can catch the tortoise. The way to escape is to diagnose the trouble in Zeno's reasoning and set it right in a general manner so we don't get caught up in new paradoxes involving the same concepts. Today they're considered to be paradoxes about the microstructure of space and time.

Naomi: OK, I understand that. What about the other two paradoxes?

John: The Dichotomy Paradox is very similar. Think of the tortoise as sitting still on the goal line. Suppose it would take the runner one minute to reach the goal traveling at a constant speed. But Zeno asks us to relook at this trip in a different manner. It will take a half minute to reach halfway there, an additional quarter minute longer to reach half the remaining distance, then an eighth minute longer to reach half of the new remaining distance, and so on. There's no end to these goals, so the final goal line is never reached.

Naomi: Just like in the Achilles Paradox, he's finding an infinite structure within a finite one and then using it to reason to a contradiction. OK, how about the last paradox?

John: The Arrow. Think of time as composed of moments, and think of motion as a series of still photos taken at those moments. At each moment a moving arrow occupies a space equal to itself. But this space doesn't move; it's just there, so the arrow is just there, too, motionless at that moment in the photo. The same holds for any other moment during the so-called flight of the arrow. So the arrow is motionless at all moments and can never move. By analogous reasoning, nothing else moves either. Therefore, motion is impossible, assuming time is composed of moments.

Naomi: I think I see the problem. If he'd taken a real photo of the arrow, he'd have seen it's blurry.

John: He didn't have a camera.

Naomi: I was joking. . . . Anyway, what are the solutions you talked about?

John: There are two major ones—Aristotle's and what is sometimes called the standard solution. One student in class today suggested that we can solve the Achilles Paradox by saying there are biological limitations on how small a step Achilles can take, so Zeno's argument fails. Mehlberg defended Zeno and said that Achilles' feet aren't obligated to obey biological limitations because Zeno is giving a theoretical argument. Very small steps are OK. It will be less misleading to think of them as movements rather than steps.

Naomi: Yeah, the comment about biology isn't a serious solution to the paradox.

John: Aristotle treats the Achilles and the Dichotomy Paradoxes by accusing Zeno of making two mistakes. First, Zeno shouldn't think of his moments as being indivisible. Second, he's making the mistake of conceiving of the continuous path taken by runners as being composed of a *completed* infinite aggregate of subpaths that exist all at once. Aristotle rejects completed infinities or actual infinities. Aristotle would have said the set of places visited by Achilles is potentially infinite but not actually infinite. Zeno considered the whole path as dependent on, or composed of, the subpaths to all these places. That's the key mistake, says Aristotle. Instead, the whole path is there independently, and then the analyst should abstract from this by a process of potentially dividing the whole into its parts. There's no upper limit on the number of these imaginary operations of division, but at no time has a completed infinity of subpaths been created, yet Zeno needs this completed infinity in order to produce his paradox. If we agree with Aristotle to reject completed infinities and stick with potential infinites, we have a way out of the Achilles and Dichotomy Paradoxes.

Naomi: So that's Aristotle's solution to the first two paradoxes?

John: Yes. His solution to the Arrow Paradox is similar. He accuses Zeno of grounding his reasoning on the false assumption that time is composed of indivisible or instantaneous moments. Aristotle says Zeno is failing to realize that duration can be divided only potentially and that means any division at any specific time produces only finitely many intervals, none of which have zero duration. If there's a finite duration, then there's time for the arrow to have different locations at different times, and thus to fly. Zeno's Arrow Paradox needs more than just a potential infinity of intervals, he said. It needs those instantaneous moments that are reached by a completed infinity of divisions of a time interval. That way Zeno can say the arrow doesn't move during the moment. The standard solution agrees with Zeno that time *can* be composed of indivisible moments or instants and implies that Aristotle has misdiagnosed the source of the error in the Arrow Paradox. But being the most influential ancient Greek in so many other areas, Aristotle's treatment of the paradoxes was accepted by almost all the experts for two thousand years. Slowly over the centuries his way out lost ground to the standard solution.

Naomi: What's the standard solution?

John: It wasn't perfected until the early twentieth century, and it approaches the paradoxes by supposing that the mathematical system of calculus that has been found to be successful for a vast range of problems throughout science is also the right system to apply in dealing with the paradoxes. The solution makes any one-dimensional spatial interval or temporal interval be a continuum in the modern sense, and this means assigning real numbers to distances and durations, not simply rational numbers. This solution agrees with Aristotle that you get a finite answer when you add up all the distances and durations for the runners, and that's what you need to get Achilles to the goal in time. But Aristotle gets his finite sum by adding up a finite number of durations taken in a potential infinity of intervals, and the standard solution doesn't. It gets its finite sum by adding up a completed infinity. It rejects the old Greek mathematics for a new one that allows a completed infinity of numbers to add up to a finite sum.

Naomi: Does the standard solution say something similar about the Arrow Paradox?

John: Yes. It implies that at a point place, an arrow can either be at rest or in motion, and which one it is depends on what happens at *other* nearby point places. The standard solution is precise about this by saying the arrow is in motion at a certain place if its derivative of position with respect to time at that place isn't zero. So the arrow can be in motion at an instant even though a single instant all by itself doesn't last long enough for motion to occur during the instant.

Naomi: Yeah, this is the way out of the paradox. Calculus gives us a new concept of motion. And with the Dichotomy, I see why calculus could help solve the problem where the runner has an infinite number of tasks of duration $1/2$ and then $1/4$ and so on. Calculus implies $1/2 + 1/4 + 1/8 + \ldots$ equals one, so the worry about having enough time to complete the tasks is gone. Completed infinites are completely legitimate in calculus. Other than out of respect for Aristotle, why do you suppose so many people once believed using potential infinities is the right way out?

John: Some people still say it's the right way out. I'm one of those people. Potential infinities are how we intuitively think about infinities, so Aristotle's solution doesn't violate our common sense the way the standard solution does. For example, consider our language. Does it contain an actual infinity of sentences? No, because nobody's ever spoken that many sentences. Nobody could. Who has the time? But for any number of sentences that have been spoken, there's potentially another sentence that could be spoken next, if only because you can use the word "and" to build a longer sentence out of two previous sentences. So our language is *potentially* infinite. The same goes for the number of tasks that a runner performs. The same ought to be said about the number of moments in any temporal interval.

Naomi: I don't see any harm in assuming that English contains an actual infinity of sentences. We can simply say that no actual speaker has uttered more than a finite number of them, and that with more time there's always another one that could be uttered.

John: But then you've got ideal English or theoretical English, not experienced English.

Naomi: I'd say we ought to be flexible about what the definition of English is.

John: I see the point you're making, but I still disagree. Back to Zeno, I think adopting all this powerful machinery of calculus and rejecting Greek common sense is too high a price to pay. First of all, I can't accept that a runner can complete an actually infinite number of tasks, no matter how much smaller each task becomes. We can complete only a potentially infinite number, a number that grows without an upper limit but is finite at any moment. I'm not talking about the runner getting tired or thirsty and needing to stop for a while, or anything like that.

Naomi: Think of a tennis ball that bounces halfway back up when you drop it. Ideally it will keep bouncing after you drop it. It will bounce back 1/2 as high, then 1/4 as high, then 1/8 as high and so on. Can't you imagine it eventually coming to a halt after a while, having had as many bounces as there are positive integers? That would be a completed infinity of bounces.

John: Yes, I can imagine it for an ideal ball, but not for a real ball. That's my point. No real ball is ideal. A real ball comes to a stop after a finite number of bounces, not after an infinite number. Stick to common sense. You wouldn't have to stray from common sense if you'd just agree with Aristotle that Achilles completes only a *potential* infinity of subtasks. That's my first complaint with the price paid for adopting the standard solution. Second, you've got to quit saying that events and their durations are infinitely divisible in your sense of infinite.

Naomi: Why quit? I mean, this feature of time is really useful in science. Science says time is continuous, and that means any finite duration is divisible into smaller parts, all the way down.

John: Intuitively, I don't have any problem agreeing that Achilles' run is a continuous entity; but that means to me it's a whole thing that's smooth with no gaps. The mathematical physicists' idea of smooth isn't anything like this. They think of smooth the way the mathematical line is smooth. Ha, what a joke! I mean, how can something be smooth and at the same time be made of a bunch of different points? That's the opposite of smooth; that's granular. Smooth means unbroken.

Naomi: I really don't think this is as contradictory as you make it seem. Yes, mathematical physicists have redefined the idea of smooth. Is that too much revision? We need the revision. I'm not denying the underdetermination of the modern theory by the data it's supposed to account for, but to build the most useful mathematical laws of nature we need to revise the word "smooth," and we need instantaneous slices of reality to be matched with the real numbers and these in turn to be matched with the points on the mathematician's line segment. And we need physical processes to be aggregates of point-events. Besides, last year in my physics lab I carefully counted all the points on a one-meter line segment and discovered that there is an actual infinity of them.

John: You couldn't do that!

Naomi: Just kidding. I'll admit that having an actual infinity of points in a line segment is counterintuitive.

John: It is. Point instants can't even be detected in our experience, and then there's the weird idea in the standard solution that between any two points in a continuous path in space or time, there's what Mehlberg said is an uncountable infinity of points in between those two points. I don't believe there's room to pack an uncountable infinity of instants into a second.

Naomi: What tool did you use to figure out that there's no room?

John: My intuitions, and my knowledge of what words mean.

Naomi: That's not a reliable guide.

John: Mehlberg says a key idea of the standard solution is that Zeno's fractions are a countable infinity of numbers abstracted from a preexisting uncountable infinity of real numbers. That puts too great a strain on my intuitions. The whole notion of one infinity being bigger than another violates the definition of the word "infinity."

Naomi: Your definition, not mine. These unintuitive structures of points and kinds of infinities help remove troubles in calculus. Historically, calculus contained contradictions—George Berkeley was the big critic back in the eighteenth century—but now they're gone thanks to accepting all those revisions that you complain are unintuitive. And there's another reason to accept this mathematical apparatus. If we can say runners' paths and planetary orbits and even the salinity in the ocean are physical continua, at least when viewed at a large scale bigger than the atomic scale, and if we can say physical processes such as planetary motion and salinity changes are aggregates of point-events, then we get to apply calculus to describe changes. The calculus helps us understand these things and make good predictions about how they behave. We've eaten the fruit of the calculus, and we're not about to spit it out.

John: I'm all for usefulness, but a physical theory can give a useful and fairly adequate treatment of reality without every little detail of the theory being taken literally. What about the idea of artifacts of the formulation?

Naomi: OK, but you seem not to be willing to trust any feature of the theory that deviates from common sense.

John: Look at it from this perspective. I'm sure you'd say there are an actual infinity of instants in any interval, but not an actual infinity of salinities in saltwater because eventually you reach the level of molecules. Salinity can't change if you don't change whole salt molecules, right?

Naomi: Yes.

John: So you're being arbitrary by accepting one actual infinity and not the other. How do you know when to take a detail of the theory literally and when not to?

Naomi: Well, I know not to take a feature literally when I have independent reasons for the failure of the feature at the microlevel. I have those reasons for salinity, but not for the microstructure of time and space. No actual experiment has ever indicated that there's a lower limit to the size of durations and distances.

John: Maybe this is just a sign you've been using instruments that are too crude to uncover the limits and that someday we'll bump up against atoms of time and space.

Naomi: Maybe, yes, but so far there's no reason to think so, and the usefulness of the assumption that time and space are continua is good evidence they really are. Common sense needs to change. The success of the idea that time is a continuum of point instants is shown by its ability to solve problems and suggest new areas of research without being contradicted by any experimental evidence. The attitude I'm using here is the same attitude that declares zero to exist because zero is so useful for solving the numerical equation $x + b = c$ in the special case when the two numbers b and c happen to be equal, and because having solutions to these sorts of equations is indispensable to the best scientific theories. To me, it looks like the standard solution isn't just coming up with some treatment that will undermine Zeno's reasoning—I agree Aristotle's reasoning does that, too—but it's using techniques and concepts that not only do that but also are needed in the development of a successful system of mathematics and physical science. Maybe that's why mathematicians and physicists generally agree that the standard solution is the correct solution.

John: Or maybe philosophers are less apt to leap into the abyss. Look, I'm not antiscience, but scientists can make mistakes just like anybody else.

Naomi: Scientists makes fewer mistakes because science is a group effort that's self-correcting.

John: You and I have very different opinions on this. I studied science once, learned that tomatoes have genes in them, and haven't eaten a tomato since.

Naomi: What?

John: Seriously though, do you always believe what science says about anything?

Naomi: Yes, but science changes from time to time, so I have to change my mind now and then.

John: I can imagine one good scientific theory saying there are point instants and another good scientific theory saying there aren't. What would you do then?

Naomi: Pray.

John: Are you serious?

Naomi: No. I'd probably suspend belief until the scientists work out their differences and produce a unified view of the world.

John: How can you be sure that there *is* a unified view of the world to be found?

Naomi: I can't. It's helpful to assume, but I can't be sure there is one. I think a big reason for the slow acceptance of the idea that time is a continuum was that for many centuries there really were inconsistencies in calculus. George Berkeley was one of the first critics; he complained about the inconsistent use of infinitesimals by Newton and Leibniz. The main goal for mathematicians for the next two hundred years was to consistently work out the necessary and sufficient conditions for being a continuum. In doing that, eventually they came to believe that potential infinities of numbers and arbitrarily large sets of numbers both should be considered to be subsets of a *completed* infinite set of numbers. The implication for Zeno is that the set of ever-decreasing intervals in the Achilles and Dichotomy Paradoxes is most properly treated as a subset of an *actually* infinite set of intervals. Now that we've got it right, and the contradictions are gone, the mathematicians and physicists have been won over to the standard solution, but it appears to me that some of the metaphysicians are lagging behind because they're still holding onto antique intuitions. I don't want to slaughter your intuitions. They're my intuitions, too. I just want to make a conservative revision in them.

John: I don't agree that your mathematics is indispensable. There's an alternative. One of the students in class said that instead of using your continuum mathematics, which requires real numbers for the values of lengths, durations, speeds, and all that, we could treat them with rational numbers by saying any duration is the difference in the rational numbers for the end and the start of the duration. If we did this, we'd still be able to agree with all measurements ever made. The idea is to find the smallest distance ever detected in any measurement and the smallest detected duration, and then pick the atoms of space and time to be smaller than this and to have rational numbers for their measures, not real numbers. I don't know too much about mathematics compared to you, but the idea is to replace calculus's differentials of real numbers with small differences of rational numbers. Good-bye, real numbers.

Naomi: You can't make do with rational numbers. If the side of a square is one unit, then the diagonal is the square root of two, which is irrational. The implication of your route is that squares no longer have diagonals.

We can't live with that. Diagonals are indispensable. To do science we really need the elegant differential equations of calculus. Elegance is a sign that we mathematical bloodhounds are sniffing out the truth. That's why the standard solution is the right solution.

John: I'm bothered by your argument about the diagonal of a square. Space has squares but time doesn't, so all your argument shows is that measures of space need real numbers, not that measures of time do.

Naomi: You're right. I forgot for a moment that time is only one-dimensional. But look, if space is a continuum, then why not treat time that way, too, especially since Einstein and Minkowski showed that space and time are just features of something more fundamental, namely spacetime?

John: Personally, I think the new concepts of the standard solution do too much damage, and we should back away. Aristotle is underappreciated. Scientific theories should be preserving intuitions, not attacking them. I'd start off with a pointless geometry.

Naomi: Was that a pun?

John: Yes, sorry, but I'm serious about backing away from instantaneous instants and point places.

Naomi: Do you want to preserve intuitions because these give us literal truth?

John: What I do know is that continuity is something to be understood by examining our perceptions using our sense organs, and not by adopting bizarre mathematical constructions. Your mathematics misrepresents. It's absurd to interpret an experienced continuum of space or of time as a set of individual points. Let's stick with an intuitive continuum and potential infinities. The actual infinite can't be encountered in experience. Besides, the completion of an infinite number of tasks is known to be a logical impossibility.

Naomi: It is? Who says?

John: There are some things we all just know by pure reason.

Naomi: I don't trust pure reason.

John: How about this? You should trust definitions, and we know from the definition of the word "task" that nobody can complete an infinite number of tasks. Think about what a task is. Completed tasks need starting and stopping times. They can't last for zero seconds, and something has to change between the start and end of the task.

Naomi: Are you requiring that someone intend to complete the task or else it's not a task?

John: No.

Naomi: OK, no problem so far.

John: And they need minimum times for their completion.

Naomi: Oops! I don't think so. It all depends on the task. Yes, a task we can detect with our sense organs requires some minimal duration, but that finiteness is something for psychologists to determine; it's not something to write into the definition of "task."

John: Well, we disagree about this, but I have another reason to think that Achilles doesn't perform an actual infinity of subtasks. Performing this many tasks is called performing a supertask. But here's why the whole idea of a supertask is nonsense. Forget Achilles for a second, and think about a lamp whose power switch is a toggle switch. Switch the lamp on for a half minute, then switch it off for a quarter minute, then on for an eighth minute, off for a sixteenth minute, and so on. Will the lamp be lit or dark at the end of a minute?

Naomi: I don't know. You mentioned this during our first bus ride together.

John: Nobody else does either. Because there should be an answer, but isn't, there was a mistake back in the reasoning, and the mistake was to suppose that it's logically possible to perform a supertask. So Achilles doesn't perform a supertask in catching the tortoise. Now think about the implication of saying this. Your reasoning and Zeno's reasoning both have him performing a supertask. The only way to restore logical consistency is to say Achilles performs a *potential* infinity of tasks, not an actual infinity. Hello, Aristotle! Can you hear me?

Naomi: OK, OK, I'll agree with you that this supertask with the lamp is incompletable in the sense that there's no last task that gets completed, but it *is* completable in another sense—in the sense that all the subtasks can be completed in a finite time. As you described the lamp, either way it ends up—on or off—is compatible with the infinite sequence of tasks being completed. There's no contradiction. The description of the initial situation fixes the status of the lamp at any finite stage, but not after the infinity of stages, so there's no reason for you to complain that we can't deduce whether the lamp is on at the end. It would be like complaining that a Sherlock Holmes detective story is defective because we can't deduce whether Holmes smoked his pipe after the last day mentioned in the story. Because there's no contradiction, you can't say that to escape the contradiction we have to accept Aristotle's treatment using potential infinity. But, look, I'll agree with you that the lamp is impossible, though for a different reason than yours. I assume the switch handle moves the same distance for each switching. So with faster and faster switching it eventually goes faster than the speed of light. That's the real reason why your infinity lamp is impossible—it's inconsistent with the speed limit of the theory of relativity. But this impossibility won't give you Aristotle's solution to Zeno's Paradoxes, because even though a real-life flip of a switch does take a minimum time, the subruns of Achilles don't take a minimum time. So it's OK to say Achilles performs a supertask. If you let me redesign the lamp so that switching has no minimum duration, then I wouldn't have a problem with the infinity lamp either.

John: You can only have Achilles perform a supertask if you idealize away from reality and don't talk about real runners and real lamps.

Naomi: Your attitude is so different from mine, but you do have allies. Some of the spirit of your opposition to completed infinities and the continuum persists today in the philosophy of mathematics called "constructivism." It's a neo-Aristotelian movement.

John: What do they say?

Naomi: Constructivists say that before they'll agree the continuum or some other mathematical object exists, it has to be constructed in a finite way from previously acceptable notions. For a constructivist, the only acceptable infinite sequences of constructions are *potentially* infinite ones, not actually infinite ones. That means that at any point in time only a finite number of members of the sequence have been constructed. In the twentieth century, the constructivists tried to place calculus on this sort of foundation, but they couldn't justify some of the important theorems that everyone uses in both math and science. This trouble caused many people to reject constructivism, and it severely weakened the neo-Aristotelian movement. That's why the vast majority of today's practicing mathematicians routinely use nonconstructive mathematics.

John: I wish I knew more about constructivism, but I just thought of one other reason why there might be trouble for your approach to applying math to reality. Think about the instantaneous speed of Achilles.

Naomi: When?

John: At any one of those ideal instants of yours. He can't change location during an instant since an instant doesn't last long enough, so his speed is defined only over some longer interval that includes the instant. That means his speed isn't intrinsic to him at the instant.

Naomi: OK. Is that a problem? Can't we just say the speed at a point-place is whatever the average speed is over a small interval including the point-place? That's what calculus does.

John: Well, there aren't intrinsic instantaneous velocities so you can't appeal to them in explaining why moving objects move from the present instant to some later instant. My worry is that, if there really were instants, then the complete state of Achilles at an instant should include that fact that he's moving. But his speed is defined only for the tiny neighborhood around the instant, and this includes a little of the past and a little of the future. So it isn't just features of the present that determine the future. However, if determinism is true, then past states of the world are supposed to be irrelevant to the determination of its future states.

Naomi: Yes, but determinism is probably false.

John: I agree with you, but you don't want to make it false by definition, do you?

Naomi: No, I guess not.

John: Well, you have, so you've got a problem.

Naomi: I see. Maybe we could redefine what an intrinsic property is. No, maybe we need to change the definition of determinism and make it say that the state of the world throughout an infinitesimal time interval is what fixes the state of the world at other time intervals. Hmm, this is an interesting problem.

John: My bus stop is nearly here.

Naomi: Just in time.

John: Hey, here's my phone number. Give me a call one of these days, and we'll talk some more. I'm really enjoying our conversations.

Naomi: Me, too. I've never told you this, but I used to be suspicious that metaphysicians were just navel gazers. Now I see things differently.

John: Good. Talk to you soon.

DISCUSSION QUESTIONS

1. How does Aristotle's solution to the Achilles Paradox differ from the standard solution?
2. Does English contain an actual infinity of sentences or only a potential infinity? How about the number of points in the physical line

from where you are to the nearest door? If someone disagreed with you about either of your answers, what sort of reasons would they be likely to offer?

3. Aristotle says Zeno misunderstands the composition of time in his Arrow Paradox. Aristotle believes time is not composed of indivisible moments. If not that, then what does Aristotle think time *is* composed of? How should we go about learning what time is really composed of?

4. How do John and Naomi disagree about the role of intuitions and common sense in regard to Zeno's Paradoxes? Who's right and why?

5. Can the theories of physical science tell us how things are, or are they always only an approximation to how things are, or is there no unique way things are that physical science can aim at?

6. Describe the infinity lamp. Is violating a speed limit the only serious problem with it?

7. Does Achilles really perform a supertask? If someone disagreed with you about this, what sort of reasons would they be likely to offer, and why would these be inadequate?

8. Explain why John is worried that Naomi will rule out determinism by her definition of "speed."

9. What is time?

FURTHER READING

Arntzenius, Frank. "Are there Really Instantaneous Velocities?" *The Monist* 83 (2000): 187–208.
  In exploring different responses to Zeno's Arrow Paradox, the article examines the possibility that there are no instantaneous velocities at a point, that a duration does not consist of points, and that every part of time has a nonzero size.
Benacerraf, Paul. "Tasks, Super-Tasks and the Modern Eleatics." *Journal of Philosophy* 59 (1962): 765–84.
  Argues that the description of the infinity lamp does not and need not fix the behavior of the lamp at the end of the switching.
Black, Max. "Achilles and the Tortoise." *Analysis* 11 (1951): 91–101.
  Black challenges the standard solution to the Achilles Paradox, arguing that contemporary mathematics isn't applicable to reality. He discusses infinity machines and supertasks and argues that Achilles does not perform a supertask.
Dummett, Michael. "Is Time a Continuum of Instants?" *Philosophy* 75 (2000): 497–515.
  A constructivist model of time that challenges the idea that time is composed of point instants. Difficult reading.
Grünbaum, Adolf. "Relativity and the Atomicity of Becoming." *Review of Metaphysics* 4 (1950–51): 143–86.

A defense of the treatment of time and space as being continua and of physical processes as being aggregates of point-events. Difficult reading.

Kirk, G. S., J. E. Raven, and M. Schofield, eds. *The Presocratic Philosophers: A Critical History with a Selection of Texts.* 2d ed. Cambridge: Cambridge University Press, 1983.
An excellent source in English of primary material on the pre-Socratics.

Moore, A. W. *The Infinite.* 2d ed. New York: Routledge, 2001.
Contains the history of the transition from Aristotle's treatment of infinity to the standard treatment begun by Cantor. Also discusses supertasks.

Newton-Smith, W. H. *The Structure of Time.* London: Routledge & Kegan Paul Books Ltd., 1980.
Chapter 6 discusses how to treat time as discrete or as merely dense rather than as continuous.

Resnick, Michael. "Quine and the Web of Belief." In *The Oxford Handbook of Philosophy of Mathematics and Logic,* edited by Stewart Shapiro, 412–36. Oxford: Oxford University Press, 2005.
See pages 429–32 for a clear discussion of the pragmatic indispensability argument for mathematical realism: "For given that we are justified in doing science, we are justified in using (and thus assuming the truth of) the mathematics in science, because we know of no other way of obtaining the explanatory, predictive, and technological fruits of science."

Sainsbury, Richard M. *Paradoxes.* Cambridge: Cambridge University Press, 1988.
Chapter 1 (pages 5–24) gives a clear presentation of Zeno's Paradoxes at the same level of difficulty as in this dialogue.

Salmon, Wesley C. *Space, Time and Motion: A Philosophical Introduction.* 2d ed. revised. Minneapolis: University of Minnesota Press, 1980.
Chapter 2 (pages 31–68) analyzes Zeno's Paradoxes through the lens of the standard solution without requiring a sophisticated understanding of calculus.

———. *Zeno's Paradoxes.* New York: The Bobbs-Merrill Company, Inc., 1970.
A collection of influential articles about Zeno's Paradoxes that were written in the period from 1911 to 1965. Salmon provides an excellent annotated bibliography of further readings.

Stacy, B. David. "The Story of the Hotel Ad Infinitum." http://scidiv.bcc.ctc.edu/Math/InfiniteHotel.html.
A very short, humorous story demonstrating some implications of Richard Dedekind's definition of an infinite set as one that can be put in one-to-one correspondence with a proper subset of itself.

Thomson, James F. "Tasks and Super-Tasks." *Analysis* 15 (1954): 1–13.
The source for the story of the infinity lamp. Thomson argues that, if his lamp is impossible, then it's doubtful that Achilles completes a supertask.

Vlastos, Gregory. "Zeno of Elea." In *The Encyclopedia of Philosophy* 8, edited by Paul Edwards, 369–79. New York: Macmillan Publishing Co., Inc., and The Free Press, 1967.
An account of Zeno's Paradoxes with careful attention to source material and to how interpretations of the paradoxes have been constructed.

Glossary

Absolute theory. *See* substantivalist theory of spacetime.

Backward causation: Causation in which the effect occurs before its cause instead of after.

Big bang theory: A generally accepted scientific theory of the origin and development of the universe, namely that it expanded from a hot, dense, microscopic, exploding fireball billions of years ago, and as it expanded it cooled to the universe we live in today.

Bilking: Acting to prevent a predicted event from occurring.

Block universe theory: Metaphysical theory that implies all of the past, present, and future is real. The name derives from the fact that a Minkowski diagram would represent events as points in a block if space and time were to be finite in all directions. Also called "eternalism."

Correlated: Two variables are correlated if they are related in the sense that systematic changes in the value of one variable are accompanied by systematic changes in the other.

Countable infinity: A set is countably infinite in size if it has the same number of members as the set of natural numbers. An uncountable infinity is not countable, and so is bigger. The set of real numbers is bigger than the set of integers, though both are infinite.

Determinism: The metaphysical theory that every event has a cause. Causal determinism implies later states of the universe are fixed by any complete state at a time (instants, say, or perhaps intervals no matter how small) in the past and by the laws of nature that are operating, assuming past states influence the future only through their influence on the present state. Some versions of determinism also require a complete state at a time to determine any past state.

Duration: The duration of an event is the amount of time it lasts.

Entropy: A measure of an isolated system's internal disorder or its energy that is unavailable for useful work. It is very probable that entropy will increase as time goes on. One sign of entropy increase is that, as time goes on, hot objects cool down, cold objects warm up, bouncing balls quit bouncing, metal rusts, and sounds die out.

Eternalism: The metaphysical theory that future, present, and past events are real. *See* block universe theory.

Euclidean geometry: A mathematical theory of size, shape, and position, having these two theorems, among others: (a) the sum of the measures of the three interior angles of any triangle is always 180 degrees, and (b) in a two-dimensional plane, if you are given a straight line and a point not on the line, there is exactly one other straight line passing through the point but never intersecting the original line.

Fatalism: The metaphysical doctrine that fate exists, in the sense that some events are forced to happen because of supernatural forces beyond our control. Compare with determinism.

First-order change: Having different intrinsic properties at two times. Compare with second-order change.

Growing universe theory: A metaphysical theory that the present and past are real but the future is not.

Indexical term: A term whose reference depends on the context or circumstances of its utterance. Examples are "you" and "last week."

Instantaneous event: An event that lasts for zero seconds.

Intrinsic property: A quality or feature of an entity as opposed to its relationships with other entities.

Light cone: In a Minkowski diagram, for a single point, all events causally connectible to that point will form a double cone of events, one cone in the point's past and one in the point's future.

McTaggart's A-series and B-series: His A-series of events orders them by their properties of how past they are, whether they are present, and how far in the future they are. His B-series orders the events by how much time an event happens before or after other events.

Metric: A function defining the distance between pairs of points of a space.

Minkowski diagram: A spacetime diagram is a representation of the point-events of spacetime. In a Minkowski spacetime diagram, a rectangular coordinate system is used, Einstein's special theory of relativity holds, normally the time axis is vertical, and one or two of the spatial axes are suppressed.

Modal realism: The philosophical theory that all possible worlds actually exist, though not in our world.

Non-Euclidean geometry: Similar to Euclidean geometry, but without theorems (a) and (b) above.

Objective: Not subjective.

Perdurance theory: The matephysical theory that perduring four-dimensional spatiotemporal objects are the ontologically basic objects, and every such object has a temporal part at every moment it exists. No claim need be made about what everyday terms mean, nor that the vocabulary of perdurance theory must be used for analyzing talk of three-dimensional objects that continue through time.

Perdure: Perduring objects cannot wholly exist at an instant. The theory of relativity treats perduring events in spacetime as the basic components of reality.

Physical time: The time that clocks are designed to measure. Compare with psychological time.

Physical possibility: A possibility that isn't ruled out by physical laws.

Possible world: A way things could be. A possible world is an entire universe, which, for an eternalist, would include all of its past, present, and future. Physically possible worlds can differ from our world, but they must have the same physical laws, so in such a world your mother cannot be a tree, but she can have been elected prime minister of England yesterday.

Potential infinity: A set is potentially infinite over time if at any time, there is a later time at which it is larger. A set is actually infinite if it contains at least as many members as there are natural numbers (positive integers). Note the reliance on time. Aristotle would have said the set of places visited by Achilles is potentially infinite but not actually infinite.

Presentism: Philosophical theory that implies only what is present is real.

Principle of the identity of indiscernibles: Leibniz's metaphysical principle that if there is no discernible difference between two objects (or situations), then they're actually identical.

Proper time: An object's proper time is the time that would be measured by a clock infinitesimally close to the object. Also called "personal time."

Psychological time: Awareness of physical time.

Quantum mechanics: A scientific theory of atomic phenomena implying that energy, momentum, charge, emitted frequencies, and so forth can have only discrete values; that is, there is a jump from one value to the next. In classical mechanics, they have a continuum of possible values.

Radiocarbon dating: A method of dating when an organism died by measuring the ratio of its ordinary carbon to its radioactive carbon;

the latter isotope will have decayed since the organism last actively ingested carbon.

Real number: Any number that is equal to a decimal number. The square root of negative one is a number that is not a real number.

Realism and anti-realism: Realism is the metaphysical theory that many facts are not subjective and that there exists an objective world external to mind. Anti-realism disagrees with realism.

Reference frame: The device that assigns locations to all points. Reference frames are normally specified using a coordinate system whose origin is fixed to some specific object. In a frame fixed to the airplane you are flying in, the plane's seats are not moving, but the Eiffel Tower is.

Relationism: A metaphysical theory that without events there would be no time, and without objects there would be no space. Also called "relationalism." Contrast with the substantivalist theory of spacetime.

Second-order change: The changes events undergo when they lose an amount of futureness, become present, or move farther into the past. Compare with first-order change.

Spacetime: The arena that events occur in. Special relativity implies that in two reference frames spacetime will divide up differently into its space and its time part if the frames are moving relative to each other.

Standard clock: The clock that any other clock must be synchronized to in order to be accurate.

Subjective: Having to do with a being's inner mental life. Compare with objective.

Substantivalist theory of spacetime: A theory that implies spacetime is a substance supplying the objective locations of all possible physical objects and their events—regardless of whether spacetime actually contains any physical objects. Also called "substantivalism." This book does not distinguish the substantivalist theory from the absolutist theory. Opposed to relationism.

Supertask: To perform a supertask is to perform an actual infinity of tasks in a finite time.

Surreal art: Art that presents impossible, unnatural, or incongruous combinations of objects.

Temporal part: A temporal stage or time-slice of a perduring object. The parts can be instantaneous. In traditional ontology, a three-dimensional object such as a fence post endures through time (in the sense of wholly existing at various times), but it has no temporal parts.

Tensed fact: A fact needed to make a tensed statement be true. Insofar as reality is all the facts, the issue is whether tense is an objective feature of reality or only a feature of how we represent reality.

Tensed theory of time: The philosophical theory implying that the concepts of past, present, and future are not subjective and also cannot be defined or explained or have their truth conditions specified in terms of the concepts of happens-before and happens-at-the-same-time-as.

Theory of relativity: Einstein's special and general theories of space and time designed to explain large-scale phenomena and high-speed phenomena. The special theory doesn't allow objects to accelerate or be affected by gravity. The general theory is free of these restrictions.

Time dilation: The stretching of time. For example, the six-month space trip, as measured by your twin's moving clock on the spaceship, will stretch to five years on your clock back on Earth. Or, looked at the other way, her moving clock runs very slowly compared to yours.

Truth conditions: The truth conditions of the statement, "Snow is white, and the moon is also," are that two conditions must hold, namely snow being white and the moon being white.

World line: The path of an object through spacetime. The world line will be a line (rather than a tube) only if at any instant the enduring object is no bigger than a point.

About the Author

Bradley Dowden is a professor of philosophy at California State University Sacremento. He is the author of the philosophy textbook *Logical Reasoning* (Thomas Wadsworth) and has been a general editor of the *Internet Encyclopedia of Philosophy* (peer-reviewed) since 1999. He is the author of encyclopedia articles on "Time," "Zeno's Paradoxes," "The Liar Paradox," "Truth," and "Fallacies," and he holds a Ph.D. degree in philosophy from Stanford University and a master's degree in physics from Ohio State University.

Made in the USA
Lexington, KY
30 August 2012